THE MATERIAL IMAGINATION

Ashgate Studies in Architecture Series

SERIES EDITOR: EAMONN CANNIFFE, MANCHESTER SCHOOL OF ARCHITECTURE,
MANCHESTER METROPOLITAN UNIVERSITY, UK

The discipline of Architecture is undergoing subtle transformation as design awareness permeates our visually dominated culture. Technological change, the search for sustainability and debates around the value of place and meaning of the architectural gesture are aspects which will affect the cities we inhabit. This series seeks to address such topics, both theoretically and in practice, through the publication of high quality original research, written and visual.

Other titles in this series

From Formalism to Weak Form: The Architecture
and Philosophy of Peter Eisenman
Stefano Corbo
ISBN 978 1 4724 4314 4

Suspending Modernity: The Architecture of Franco Albini
Kay Bea Jones
ISBN 978 1 4724 2728 1

The Architecture of Industry
Changing Paradigms in Industrial Building and Planning
Edited by Mathew Aitchison
ISBN 978 1 4724 3299 5

Architecture in an Age of Uncertainty
Edited by Benjamin Flowers
ISBN 978 1 4094 4575 3

Charles Robert Cockerell, Architect in Time
Reflections around Anachronistic Drawings
Anne Bordeleau
ISBN 978 1 4094 5369 7

Forthcoming titles in this series

In-Between: Architectural Drawing and Imaginative Knowledge
in Islamic and Western Traditions
Hooman Koliji
ISBN 978 1 4724 3868 3

Phenomenologies of the City
Studies in the History and Philosophy of Architecture
Edited by Henriette Steiner and Maximilian Sternberg
ISBN 978 1 4094 5479 3

The Material Imagination

Reveries on Architecture and Matter

Edited by
Matthew Mindrup
University of Canberra, Australia

Routledge
Taylor & Francis Group
LONDON AND NEW YORK

First published 2015 by Ashgate Publishing

2 Park Square, Milton Park, Abingdon, Oxfordshire OX14 4RN
711 Third Avenue, New York, NY 10017

Routledge is an imprint of the Taylor & Francis Group, an informa business

First issued in paperback 2017

British Library Cataloguing in Publication Data
A catalogue record for this book is available from the British Library

Library of Congress Cataloging-in-Publication Data
The material imagination : reveries on architecture and matter / [edited] by Matthew Mindrup.
     pages cm. -- (Ashgate studies in architecture)
   Includes bibliographical references and index.
   ISBN 978-1-4724-2458-7 (hbk)
   1. Architecture--Philosophy. 2. Matter. I. Mindrup, Matthew, editor.
   NA2500.M37 2014
   720.1--dc23

                                                          2014023407

ISBN 978-1-4724-2458-7 (hbk)
ISBN 978-1-138-57351-2 (pbk)

Contents

List of Figures

About the Editor

Matthew Mindrup is Assistant Professor of Architecture at the University of Canberra. An architect by training, Matthew completed a Ph.D. in Architecture and Design at Virginia Tech University in 2007 on the physical and metaphysical coalition of two architectural models assembled by Kurt Schwitters in the early 1920s. Matthew's ongoing research in the history and theory of architectural design locates and projects the implications that materials have in the design process. He has presented some of this research at conferences and published others in *Journal of Architectural Education* (JAE) and *Time & Architecture* (T&A). Today he lives in Canberra, Australia, where he teaches courses in architectural design, technology and the history of architectural theory.

About the Contributors

Manuela Antoniu obtained her professional (B.Arch.) and post-professional (M.Arch.) degrees in Canada, and her doctorate at the Architectural Association in London. Her work has been exhibited in Europe and North America. She has published with, among others, Princeton Architectural Press, McGill-Queen's University Press, and Routledge (for *Architectural Theory Review*). Having most recently taught at The Bartlett School of Architecture (University College London) in History and Theory, she is currently undergoing rigorous Zen training in a Buddhist temple in Japan.

Alessandro Ayuso is Senior Lecturer at the University of Westminster's School of Architecture and the Built Environment in London, UK. Alessandro is also currently completing a Ph.D. by Design thesis at the Bartlett School of Architecture, entitled "Body Agents: Figures as Agents for Design." Alessandro received his Bachelor of Architecture from Virginia Tech at Blacksburg, Virginia, and studied urban design as Fellow in Syracuse University's M.Arch. II Program in Florence, Italy. He has taught, lectured, and sat as a critic at numerous universities. He has also worked in professional design practices in New York City, including as a partner at MAKE Design. He has recently co-founded the Agency for Interior Urbanism, a platform for investigating alternative strategies for urban design.

Carolina Dayer is currently a Ph.D. candidate at Virginia Tech University's Washington–Alexandria Architecture Center (WAAC) in Alexandria, VA where she is also Lecturer and Consortium Program Coordinator for CalPoly University, San Luis Obispo, CA. She is a licensed architect in her native country, Argentina. Her research, teaching, and original work center on investigations and demonstrations of the role of the imagination and the reality of everyday life in architectural drawing. She has composed and instructed several theory and history courses and design studios utilizing explorations through drawing and making to pursue theoretical questions of the material and magical imagination. Her doctoral work focuses on the intersections between the literary practice of magic realism and the material imagination in architectural drawing. Additionally, Carolina has published, lectured internationally, and organized symposia on matters of the imagination and drawing practices, collaborating with leading scholars in the field. Her personal design work has been exhibited in Argentina and in the United States.

Paul Emmons is a registered architect and Associate Professor at the Washington–Alexandria Architecture Center of Virginia Tech, where he directs the Ph.D. program in Architecture + Design. He earned a Ph.D. from the University of Pennsylvania and an M.Arch. from the University of Minnesota. His research on the history and theory of architectural practice focuses on questions of drawing and representation. This work has been presented around the world at conferences and widely published in articles and numerous book chapters. He also co-edited *The Cultural Role of Architecture: Contemporary and Historical Perspectives* (Routledge, 2012).

Ufuk Ersoy is Assistant Professor at Clemson School of Architecture. He completed his Ph.D. in Architecture at the University of Pennsylvania. In 2010, Ersoy co-edited a special issue of the journal *World Architecture* on "Architecture in Turkey: A Glocal Production," with Sebnem Yucel. Most recently, he published his essay "To See Daydreams: The Glass Utopia of Paul Scheerbart and Bruno Taut" in *Imagining and Making the World: Reconsidering Architecture and Utopia*, edited by Nathaniel Coleman, and lectured at Newcastle University, Virginia Tech, University of Maryland, and WAAC schools of Architecture. Apart from his scholarly work, in 2010, Ersoy won a prize in the Izmir Opera House Competition in collaboration with Tozkoparan Architects, Izmir and Ove Arup, London.

Jonathan Foote is currently Lecturer for CalPoly University in San Luis Obispo, CA. He completed his Ph.D. in 2013 at Virginia Tech, entitled, "Well-Tempered Building: Michelangelo's Full Scale Template Drawings at San Lorenzo." His research, teaching, and creative work focus on the process of translation between an architectural idea and its realized construction. Jonathan has published and lectured internationally on matters of fabrication theory from the Renaissance to today. Additionally, as a leader in the field of design/build education, he has conducted international, invited workshops, and has acted as a visiting critic and lecturer at numerous universities.

Jonathan Hale is an architect, Associate Professor and Reader in Architectural Theory at the University of Nottingham. Within the Department of Architecture and Built Environment he is Head of the Architectural Humanities Research Group. Research interests include: architectural theory and criticism; phenomenology and the philosophy of technology; the relationship between architecture and the body; museums and architectural exhibitions. He has published widely in these areas and has obtained grants from the Engineering and Physical Sciences Research Council (EPSRC), the Leverhulme Trust, the British Academy, and the Arts Council. He is founding Chair and current Steering Group member of the international subject network: Architectural Humanities Research Association (AHRA). He is currently working on a book for the Routledge "Thinkers for Architects" series on the work of French philosopher Maurice Merleau-Ponty.

Jonathan Hill is an architect and architectural historian, Professor of Architecture and Visual Theory, and Director of the M.Phil./Ph.D. Architectural Design program at the Bartlett School of Architecture, University College London. Jonathan is the author of *The Illegal Architect* (1998), *Actions of Architecture* (2003), *Immaterial Architecture* (2006), and *Weather Architecture* (2012), the editor of *Occupying Architecture* (1998) and *Architecture—the Subject is Matter* (2001), and co-editor of *Critical Architecture* (2007).

Steven Holl has realized numerous significant cultural, civic, academic, and residential projects both in the United States and internationally since the founding of his practice. Steven has been recognized with architecture's most prestigious awards and prizes including a 2012 AIA Gold Medal, the RIBA 2010 Jencks Award, and the first ever Arts Award of the BBVA Foundation Frontiers of Knowledge Awards (2009). Additionally, he is a tenured professor at Columbia University's Graduate School of Architecture and Planning. He has lectured and exhibited widely and has published numerous texts, including his most recently published *Horizontal Skyscraper* (2011), *Scale* (2012), and *Color Light Time* (2012).

Sandra Karina Löschke is Senior Lecturer and Director of Architecture Design Research at the University of Sydney. Her research investigates links between aesthetics, design, and technology in exhibition architecture from the 1920s to the present, focusing on how these linkages played a significant role in progressing new disciplinary paradigms and expanding the culture of architectural knowledge. Her work includes case studies, design projects, and exhibitions, and endeavors to engage theoretical and historical frameworks with the reality of contemporary architectural practice. As an architect, her projects include award-winning buildings for Foster and Partners in London and Stephan Braunfels in Munich. Her own work has been exhibited at the Venice Architecture Biennale and exhibitions in Australia and Asia. She studied architecture at the Bartlett/UCL, the Architectural Association London, and the University of New South Wales. Sandra is currently working on an edited book entitled *Materiality in Architecture* (Routledge).

John Roberts is Lecturer in Architecture at the School of Architecture and Built Environment, University of Newcastle, NSW, Australia. He teaches Architectural Design, Site Studies, and Communications in the Undergraduate program, and lectures in Site Studies and supervises Masters' and postgraduate research. His M.Phil.(Arch.) thesis considered prospect and refuge in Alvar Aalto's architecture. Recent publications have investigated housing by Aalto and Utzon, landscape in Utzon's architecture, and landscape symbolism in the Chinese garden. Current research interests include: architecture and the natural world; the color white; ruins; and landscape in architecture. He is currently doing Ph.D. research on landscape in the work of Alvar Aalto.

Nicholas Temple was born in Australia and is currently Professor of Architecture at the University of Huddersfield in the UK, having previously been Professor and Head of the School of Architecture at the University of Lincoln. A graduate of Cambridge University, Temple studied architecture under Dalibor Vesely, Peter Carl, and Joseph Rykwert and later won the Rome Scholarship in Architecture (1986–88) at the British School at Rome. He completed a Ph.D. in 2000 on architecture and urbanism in Rome during the Pontificate of Julius II (1503–13), and was recently a Paul Mellon Rome fellow in 2012. A qualified architect and Fellow of the Royal Society of Arts and the Royal Historical Society, Temple has been in both full-time and part-time architectural practice, in Cambridge, Sheffield, and Leeds. His research interests focus primarily on the history and theory of architecture, urbanism, and geography, with specific focus on Europe, having published many books and articles, including, *Disclosing Horizons: Architecture, Perspective and Redemptive Space* (Routledge, 2006), *renovatio urbis: Architecture, Urbanism and Ceremony in the Rome of Julius II* (Routledge, 2011), and having co-edited *The Humanities in Architectural Design: A Contemporary and Historical Perspective* (Routledge, 2010), *Architecture and Justice: Judicial Meanings in the Public Realm* (Ashgate, 2013) and the forthcoming *Bishop Robert Grosseteste and Lincoln Cathedral: Tracing Relationships between Medieval Concepts of Order and Built Form* (Ashgate, 2014).

Dan Willis is Professor of Architecture at Penn State University. He is a practicing architect and a past winner of the Hugh Ferriss Prize for architectural drawing. Between 2002 and 2009, he served as the Department Head of Architecture at Penn State. He is the author of *The Emerald City and Other Essays on the Architectural Imagination* (Princeton Architectural Press, 1999), and co-editor for and contributor to *Architecture and Energy: Performance and Style* (Routledge, 2013). He is currently engaged in research with the Energy Efficient Buildings Hub at the Philadelphia Navy Yard (funded by the U.S. Department of Energy), and has organized three "Architecture and Energy" symposia around that work. He continues to work on a book about craftwork and design. His writings and reviews have appeared in *Harvard Design Magazine*, and his essays are included in *Judging Architectural Value* (2007) and *Fabricating Architecture* (2010). He is also one of the founding members of Penn State's Center for Research in Design and Innovation (CRDI). His recent research focuses on the differing disciplinary understandings of "design."

Tracey Eve Winton is an architect and iconographer who holds a Ph.D. in the History and Philosophy of Architecture from Cambridge University, and an M.Arch. in the History and Theory of Architecture from McGill. She teaches Studio, Cultural History, and Urban History at the University of Waterloo School of Architecture, Canada, and in their Rome program. Every year, she and her second-year students stage an original theatrical production. She is currently completing a translation of the *Hypnerotomachia Poliphili* (1499) and a critical commentary. Her architectural research interests include urban morphology, sustainable design, craftsmanship, phenomenology, the relationship between theatre and the ideal city, landscape in Renaissance painting, the role of materials in natural magic and alchemy, history and iconography of the museum, ruins and architectural spoils, and the art of memory.

Acknowledgements

When I was a child, I read a quote from the poet Shel Silverstein that continues to resonate with me today:

> *There are no happy endings. Endings are the saddest part. So just give me a*
> *happy middle and a very happy start.*

I cannot say for certain when this book had its start. As a concept, history books attribute the beginnings for the material imagination to the French philosopher of science Gaston Bachelard. Like many other architects, Bachelard's *The Poetics of Space* was a staple of architectural education since phenomenology began to have a major impact on architectural theory in the 1970s. For myself, I owe a debt to Daniel Willis and Donald Kunze for introducing me to Bachelard's work during my undergraduate studies at the Pennsylvania State University. From Wesley Wei I developed an approach towards designing with the propensity of materials as a graduate student at the University of Pennsylvania, and Marco Frascari nurtured its study at Virginia Tech's Washington–Alexandria Architecture Center (WAAC). If anything could be considered a start, these individuals unquestionably made it a very happy one.

In preparing this anthology I enjoyed the support of my contributors. Among them I should mention Carolina Dayer, whose confident enthusiasm for the project vitalized my own pursuit of its completion. I thank Jonathan Foote, Paul Emmons and especially Donald Kunze for providing critical editorial comments on the introduction. I am indebted to my friend Nadir Lahiji for his support and mentorship navigating through the editorial process. Lastly, I am grateful to my wife Franziska for her enduring support, thoughtful comments and patience.

At Ashgate, I must extend my most sincere gratitude to Valerie Rose, who enthusiastically supported this project from the very beginning. I am grateful to her patience and help throughout this process.

Matthew Mindrup

In memory of our friend and mentor, Marco Frascari

Editor's Note
Interrogating the Gap Between the Material and Formal Imagination: An Introduction
Matthew Mindrup

> *There is a line in Dante (Purgatorio XVII.25) that reads: "Poi piovve dentro a l'alta fantasia" (then rained down in the high fantasy …). I will start this evening with an assertion: fantasy is a place where it rains.*[1]
>
> *Italo Calvino*

During a quiet evening thirty years ago, Italo Calvino sat down to write about the fecundity of fantasy in the fourth of six planned "memos" for the next millennium. Meditating on the spark of the imagination, Calvino drew upon the struggle of Dante Alighieri to describe in terms of the senses the manner by which images formed in his mind like rain falling from heaven. Confronted by the bombardment of visual images in today's world of mass consumerism, globalization, and accelerated communication, Calvino lamented the devaluation of words. Where Dante saw poetry through the substantive experience of rainfall, Calvino then evokes rainfall as a general principle, to see in what way the matter of works of art—here we include architecture—originate through the imagination of the author and their audiences.

The French philosopher of the science, Gaston Bachelard, observes how precisely this "material cause" of imaginative activity has been neglected by philosophical discourse.[2] In *Water and Dreams*, Bachelard proposes a distinction between the formal and material imagination by considering how the propensity of images arising from matter project deeper and more profound experiences than those of form.[3] In reference to Bachelard's distinction, Ivan Illich observes that "we cannot separate our experience of passion from the element of fire and cannot imagine fire without passion."[4] For Illich, our concepts of love, the hearth, rage, and passion originate in our contact with diverse matter as "stuff" that is imagined as fire. As Illich's example illustrates, the material imagination is not a mute receiver of sensory experience but "the faculty of forming images which go beyond reality, which sing reality."[5] In this singing of reality, Bachelard reserves a special place for poetry, out of which he argues "the seed of a world" can emerge.[6] Rather than the physical universe, Bachelard's "world" is akin to what Nelson Goodman describes in

Ways of Worldmaking as the cultural artifacts, systems of organization and meanings created by a group of people at any one time.[7] Through poetry, Bachelard contends that the contemplative meditation or reverie of material experience is "written."[8] Architecture, whose aim it is to structure the encounters of subjects with the "stuff" of buildings, cannot escape this matter of poetic reverie. Material experience *is* an essential condition for the imagination of architecture.

In recent years architectural discourse has witnessed a renewed interest in materiality under the guise of such familiar tropes as "material honesty," "form finding," or "digital materiality."[9] Motivated in part by the development of new materials and an increasing integration of designers in the process of fabricating architecture, a proliferation of recent publications from both practice and academia explore the pragmatics of materiality and its role as a protagonist of architectural form. Yet, as the ethos of material pragmatism gains more popularity, theorizations about the poetic imagination of architecture continue to recede. Compared to an emphasis on the design of visual form in architectural practice, the material imagination is employed when the architect "thinks matter, dreams in it, lives in it, or, in other words, materializes the imaginary."[10] As an alternative to a formal approach in architectural design, the chapters compiled in this book challenge readers to rethink the reverie of materials in architecture through an examination of historical precedent, architectural practice, literary sources, philosophical analyses, and everyday experience. Focusing on matter as the premise of an architect's imagination, each chapter identifies and graphically illustrates how material imagination defines the conceptual premises for making architecture.

These mediations are important contributions to a practice of architecture in which the formal is increasingly valued over the material. The privileging of form over matter is deeply imbedded in the drawing practices of architects today. Since as early as Leon Battista Alberti's promotion of the architect as a draftsman of orthographic drawings in his architectural treatise, *De Re Aedificatoria*, delineation has relegated materiality to a secondary role. As Alberti reasoned, design consists "in finding the correct, infallible way of joining and fitting together those lines and angles which define and enclose the surfaces of the building."[11] Conceived in the mind, these "lineaments" were embodied in the abstract language of Vitruvius's *ichonographia*, *orthographia*, and *scenographia*, the drawings that used plan, elevation, and profile to "narrate" the building as a set of projected and rotated lines.[12] For the Renaissance architect, the distance between form and matter, drawing and construction could still be resolved by the involvement of the architect in building the actual edifice. Georgio Vassari reiterates this distancing between architect and building nearly a century later:

> [Architecture's] designs are composed only of lines, which so far as the architect
> is concerned, are nothing else than the beginning and the end of his art, for all
> the rest which is carried out with the aid of models of wood formed from the said
> lines, is merely the work of carvers and masons.[13]

With the exception of some academic and professional design–build workshops, the architect remains absent from the building site today. The contemporary promise of digital fabrication to reunite the architect with the act of building through "digital craft" has an effect of desensitizing the poetic imagination of architecture by the very means it proposes to revive it.[14] This is not to say that a craftsman does not dream with their tools but that the chair behind a computer mouse and keyboard does little to diminish the distance between an architect and his/her poetic reverie of material experience in the imagination of architecture.

The split between form and matter that finds its parallels in architectural practice has a long philosophical tradition. In Aristotle's concept of hylomorphism, *hyle* (matter) is given shape by *morphe* (form).[15] Matter is presented as an inert, undifferentiated servant of form that gives it its presence. As the form-giver of buildings, Aristotle endows his architect with an imagination that resides between sensible and rational experience as a faculty that forms images.[16] Yet, as Bachelard observes, imagination is really our ability to deform what we perceive.[17] If there is no change, no unexpected combination of images, there is no imaginative act. There is only perception or the memory of a perception. Further, by characterizing matter as that which is given form, Aristotle relegates the resistances of matter to a secondary status. Certainly, already in the corbelling of Neolithic dolmens, one finds evidence for an imagination that permits architects to *dream with matter*, not as a surrogate to form but as the *sine qua non* for its being.

The philosophy of Giambattista Vico identifies the existence of a material imagination at the beginning of human understanding. In a discussion about the invention of language, Vico proposes a theory of epistemology emerging from the imagination of sensible experience. Best known for his critique of René Descartes through the principle of *"verum factum,"* Vico reasons that "the true" is verified by creative invention and not the rational interrogation of observation.[18] Vico speculates that, at the beginning of language, the first humans must have had minds resembling children imitating objects with mute gestures and monosyllabic cries.[19] When thunder struck, the first people perceived thunder as a voice, unaware that they had projected their own shouting to the shaking sky, animating nature with their own human qualities and actions. Jove was thus born as the name for thunder and for the first time the human spirit had before it a given: an imaginative universal.[20] This ability of consciousness to give form to fear caused by the thunderous sky is achieved by the creative power of *fantasia* to produce identity.[21] With this power, the mind achieved what Vico calls *sapienza poetica* (poetic wisdom).[22] Because of this creative power, Vico argues that human language is essentially metaphorical, not in Aristotle's view of metaphor as a likeness or similarity, but as an imaginative act in which objects of experience are brought into being.[23]

Bachelard is similarly critical of an Aristotelian understanding of metaphor in *Poetics of Space*, preferring an approach to the material imagination that recalls Vico's concept of *fantasia*. Bachelard classifies metaphor as a figure of speech that "gives concrete substance to an impression that is difficult to express."[24] For Bachelard this form of metaphor is a "false image," not having "the direct virtue of an image formed in spoken reverie."[25] As a "fabricated image" created by mere comparison Bachelard reasons that metaphor is different from the "deep, true, genuine roots" of an image emerging from the contemplative formation of spoken words.[26] To illustrate his point, Bachelard uses the example of a house, suggesting: "Of course, thanks to the house, a great many of our memories are housed, and if the house is a bit elaborate, if it has a cellar and a garret, nooks and corridors, our memories have refuges that are all the more clearly delineated."[27] The paradox of Bachelard's critique of metaphor is his evocation of a material image, a house, to use in the fabrication of a metaphor for different states of the human mind.[28] Bachelard's poetic reverie is not Vico's *fantasia* but the efficacy of an image that emerges from "deep, true, genuine roots," and its validity in the creation of a metaphor is comparable to, and particularly strong in, the creation of architecture.

Towns and buildings, as well as other man-made objects, domesticate the physical and socio-political world, but they also structure our experiences and mediate our relationship to it. By externalizing mental structures and images that we develop from our confrontation with the world, buildings are what the Finnish architect and critic, Juhani Pallasmaa, has described as "spatial and material metaphors."[29] This role of architecture is made particularly clear in Le Corbusier's design for the roof terrace of the Beistegui Apartment (1929–31). Employing the surrealist game of Le cadaver exquis (exquisite corpse) as an architectural design strategy, the meeting between a floor, wall and ceiling plane of a room are analogous to the fold of paper planes that unite the head, body and feet of a drawn "exquisite corpse." Corbusier designs the entire terrace to exploit the coincidence of the elements that define interior and exterior space at the "folds" of the architecture such that a lawn meets the parapet wall as a carpet, which meets the sky as an interior room with a fireplace. Yet, as an exquisite corpse, Le Corbusier's design does not propose that architecture is a material and spatial metaphor for lived experience by exchanging the lawn for a carpet or the empty sky for a ceiling. Rather, it describes how an architect's sensible experience of the floor, wall, or ceiling is used to imagine a truly novel architecture in a field of circulating meanings between interior and exterior or above and below. Certainly it was because of a similar recirculation of meaning that Calvino used "rainfall," not to reiterate Dante's metaphor for poetry but for the removal of weight from Lucretian atoms.[30]

The challenge of constructing embodied buildings is complicated by the contemporary role of an architect, who must conceive them *in absentia*. Since the architect today typically sits at a drafting table, computer screen, or cutting board, the drawing or model becomes the immediate medium for projecting future constructions. As David Leatherbarrow observes, this situation is uniquely different from the other arts, arguing that:

> *Poets make poems, painters make paintings, and musicians music. Architects, however, do not make architecture; they make drawings and models of it—representations meant to direct the development of something conceived into something constructed.*[31]

In the same spirit, Marco Frascari developed his critique for the use of drawings in the design process. Taking his cue from historical precedent, Frascari reasoned that the drafting of lines in an architectural drawing has its beginnings as re-enactment of the taut lines craftsmen use to locate materials and their assembly on a construction site.[32]

In a reference to recent studies in neuroscience—that suggest how the same neural and cognitive processes permitting architects to perceive and move through space are also the basis for their conceptual systems—Frascari argues, "we make buildings and they make us."[33] For Frascari, this explains how a graphic process that an architect uses to construct a drawing of architecture can also act as an invitation for the construal of their constructive possibilities:

> *Any architectural drawing is not just an aggregate of arbitrary signs that stands for something else—two lines make a wall, dash lines indicate something hidden, and so on—but they bring together signs that derive their meanings from the embodying of their tracing into the events that they represent.*[34]

In creating these "genetic representations" of architecture, Frascari encourages an architect's reverie of the different ways that materials are used to create drawings by exploring specific physical events of construction and inhabitation.[35] At the drafting board, our embodied memories become material in the imagination of architecture.

Through their encounter with materials, architects exploit precisely the incipient tendencies and propensities of matter that elude explication. Jane Bennett proposes the existence of a "gap between concept and reality, object and thing" in which "inanimate things have a life, that deep within is an inexplicable vitality or energy, a moment of independence from and resistance to us and other bodies."[36] This "thing-power" of materials is for Bennett a vibrant, vital resistance to the formative desires of an artisan detached from matter.[37] It is certainly an experience of matter with which an architect, cook, or anyone who participates in the transformation of materials is familiar. For an architect in particular, graphite marks on paper, bent card, and found modeling materials are not simply inert but endowed with a vital propensity that invites the formative imagination to use them in the conception of architecture.

László Moholy-Nagy, the second director of the *Vorkurs* (Preliminary Course) at the early-twentieth-century German Bauhaus was deeply concerned with developing his students' appreciation for the unique qualities of materials in architecture. For Moholy-Nagy the principal task of an architect's earliest studies is a sharpening of their perceptiveness—to become more "aware"—of the propensities of different materials. The Bauhaus intended students to acquire a "grasp of materials through actual experience … such as is never attained through book knowledge in the usual school exercises and the traditional courses of instruction."[38] Inspired by the Hungarian-born botanist and nature philosopher Raoul Francé's proposal for "using nature as a constructional model in creative technique,"[39] Moholy-Nagy's aim was for students to translate the productive principles of materials to the creative imagination of functional form in different media and—as the original German title, *Von Material zu Architektur* of his book *The New Vision*, suggests—architecture. This approach sought to encourage young architects to think with materials by inserting them precisely at the gap between the material and formative imagination. It is in these exchanges between material and form that the architect engages in the poetic reverie of architecture.

Written by professionals and academics, the chapters in this book meditate on the imagination of materials in architecture. They are divided into three groups, bundling differently styled reflections on how material imagination figures in reveries on the fabrication, conception, and perception of architecture. As a catalog of thought, the collection argues that the design of architecture resides in the material imagination.

The first group of chapters re-examines the role of diverse materials and methods employed in the conception of architecture. Similar to the work of a craftsman, an architect in reverie collaborates with his material. When sketching or modeling an imagined place, the pencil, pen, or knife creates a direct haptic connection between the object, its representation, and the designer. As an architect sketches the form of an object, he or she touches and feels its surface and internalizes its character. In the gap between material and architectural form, the propensity of an object that is held in the palm of the hand depends upon how well material corroborates with the architect's imagination of form. Chapters in this section explore how the image of architecture is a bridge between an imagining mind and material experience.

Authors grouped in the second section of this book explore historical and contemporary reveries inspired by or resulting from the act of assembling materials into places. When an architect–craftsman stacks stones or polishes hard metal, the activity becomes vividly lodged in their memories. Furthermore, in working with stone, iron, or even plastic, the will of the architect who wishes to fabricate a thing must accommodate materials' properties. As Adolf Loos reasons in a story he invented about a stonemason's failed attempt to craft an exact copy of a salamander in stone, although he had worked with his materials and tools his entire life, he developed an "eye" for them that resisted the imposition of form, encouraging the stonemason to think *with* his materials rather than *about* them.[40] Equally, an architect who specifies materials must be endowed with this kind of material foresight. The chapters in this section expose the perceptual awareness of the architect's particular constructive transformation of materials as a form of reverie whereby the immaterial comes to resonate in the material and *vice versa*.

The book concludes with investigations of the different ways sensory experience is organized, identified, and/or interpreted with respect to the material imagination. These chapters are concerned with the weaving of experience and imagination considering how the architect's perception measures and imagines things within a context formed by materiality and the body. By entering a space, an unconscious exchange occurs; the architect enters the space, but the space also enters and occupies them. Vividly interwoven in an architect's sensory experience, material imagination is a source through which the making of architecture is engendered with meaning and significance. What permitted Ruskin's lamp to cast forth its light into the darkness of nineteenth-century industrial craft remains for the contemporary architect in an expanding field of new material technologies and methods of fabrication: the architect's imagination resides in a gap between matter and form, where their corroboration with materials inspires architectural images.

BIBLIOGRAPHY

Adriaenssens, Sigrid, et al. (eds.), *Shell Structures for Architecture: Form Finding and Optimization* (London: Routledge, 2014).

Alberti, Leon Battista, *On the Art of Building in Ten Books*, trans. J. Rykwert et al. (Cambridge, MA: MIT Press, 1991).

Aristotle, *Aristotle in 23 Volumes*, vol. 23, trans. W.H. Fyfe (Cambridge, MA: Harvard University Press; London: William Heinemann Ltd., 1932).

Aristotle, *De Anima*, trans. D.W. Hamlyn (New York, NY: Oxford University Press, 1968, 1993).

Aristotle, *De Anima*, Book II, trans. R.D. Hicks (Cambridge, England: Cambridge University Press, 1907).

Bachelard, Gaston, *Air and Dreams: An Essay on the Imagination of Movement*, trans. E.R. Farrell and C.F. Farrell (1988; Dallas, TX: Dallas Institute Publications, 2002).

Bachelard, Gaston, *The Poetics of Reverie: Childhood, Language, and the Cosmos* (Boston, MA: Beacon Press, 1971).

Bachelard, Gaston, *Poetics of Space* (Boston, MA: Beacon Press, 1994).

Bachelard, Gaston, *Water and Dreams*, trans. E.R. Farrell (Dallas, TX: Pegasus Foundation, Dallas Institute of Humanities and Culture, 1999).

Bennett, Jane, *Vibrant Matter: A Political Ecology of Things* (Durham, NC: Duke University Press, 2010).

Calvino, Italo, *Six Memos for the Next Millennium* (New York, NY: Vintage Books, 1993).

Frascari, Marco, *Eleven Exercises in the Art of Architectural Drawing: Slow Food for the Architect's Imagination* (Abingdon, Oxon; New York, NY: Routledge, 2011).

Goodman, Nelson, *Ways of Worldmaking* (Indianapolis, IN: Hackett Publishing Co., 1978).

Gramazio, Fabio and Kohler, Matthias, *Digital Materiality in Architecture* (Baden: Lars Müller Publishers, 2008).

Illich, Ivan, *H_2O and the Waters of Forgetfulness* (Dallas, TX: Dallas Institute of Humanities & Culture, 1985).

Kolarevic, Branko, "The (Risky) Craft of Digital Making," in *Manufacturing Material Effects: Rethinking Design and Making in Architecture*, eds. B. Kolarevic and K. Klinger (New York, NY and London: Routledge, 2008), pp. 119–28.

Leatherbarrow, David, *The Roots of Architectural Invention: Site, Enclosure, Materials* (Cambridge, England: Cambridge University Press, 1993).

Leatherbarrow, David, "Showing What Otherwise Hides Itself: On Architectural Representation," *Harvard Design Magazine* (Fall 1998): pp. 50–55.

Loos, Adolf, *Ins Leere Gesprochen* (1921; Vienna: Prachner, 1981).

Moholy-Nagy, László, *The New Vision: Fundamentals of Bauhaus Design, Painting, Sculpture, and Architecture*, trans. D.M. Hoffmann (New York, NY: Dover Publications, 2012).

Pallasmaa, Juhani, *The Embodied Image: Imagination and Imagery in Architecture* (Chichester: Wiley, 2011).

Schröpfer, Thomas, *Material Design: Informing Architecture by Materiality* (Basel: Birkhäuser, 2011).

Vasari, Georgio, *Vasari on Technique*, trans. L.S. Maclehose, ed. G.B. Brown (London: J.M. Dent and Co., 1907).

Vico, Giambattista, *The New Science of Giambattista Vico*, trans. T.G. Bergin and M.H. Fisch (Ithaca, NY and London: Cornell University Press, 1968).

Vico, Giambattista, *On the Most Ancient Wisdom of the Italians: Unearthed from the Origins of the Latin Language*, trans. L.M. Palmer (Ithaca, NY: Cornell University Press, 1988).

NOTES

1   Italo Calvino, "Visibility," in *Six Memos for the Next Millennium* (New York, NY, 1993), p. 81.

2   Gaston Bachelard, *Water and Dreams*, trans. E.R. Farrell (Dallas, TX, 1999), p. 2.

3   Ibid., p. 1.

4   Ivan Illich, *H_2O and the Waters of Forgetfulness* (Dallas, TX, 1985), p. 6

5   Bachelard, *Water and Dreams*, p. 16.

6   Gaston Bachelard, *The Poetics of Reverie: Childhood, Language, and the Cosmos* (Boston, MA, 1971), p. 1, 6.

7   Nelson Goodman, *Ways of Worldmaking* (Indianapolis, IN, 1978).

8   Ibid., p. 6.

9   For example: Thomas Schröpfer, *Material Design: Informing Architecture by Materiality* (Basel, 2011); *Shell Structures for Architecture: Form Finding and Optimization*, eds. Sigrid Adriaenssens et al. (London, 2014); Fabio Gramazio and Matthias Kohler, *Digital Materiality in Architecture* (Baden, 2008).

10   Gaston Bachelard, *Air and Dreams: An Essay on the Imagination of Movement*, trans. E.R. Farrell and C.F. Farrell (1988; Dallas, TX, 2002), p. 7.

11   Leon Battista Alberti, *On the Art of Building in Ten Books*, trans. J. Rykwert et al. (Cambridge, MA, 1991), p. 7.

12   Ibid., p. 34.

13   Georgio Vasari, *Vasari on Technique*, trans. L.S. Maclehose, ed. G.B. Brown (London, 1907), pp. 206–7.

14   Branko Kolarevic, "The (Risky) Craft of Digital Making," in *Manufacturing Material Effects: Rethinking Design and Making in Architecture*, eds. B. Kolarevic and K. Klinger (New York, NY and London, 2008), pp. 119–28.

15   Aristotle, *De Anima*, Book II, trans. R.D. Hicks (Cambridge, England, 2007) (412a 15–16).

16   Aristotle identifies it as "that in virtue of which an image occurs in us." Aristotle, *De Anima*, trans. D.W. Hamlyn (New York, NY, 1968, 1993), Book III (428aa1–2).

17   Bachelard, *Air and Dreams*, p. 1.

18   Giambattista Vico, *On the Most Ancient Wisdom of the Italians: Unearthed from the Origins of the Latin Language*, trans. L.M. Palmer (Ithaca, NY, 1988), pp. 45–7.

19   Giambattista Vico, *The New Science of Giambattista Vico*, trans. T.G. Bergin and M.H. Fisch (Ithaca, NY and London, 1968), pp. 76–7 (specifically, axioms 225 and 231).

20   Ibid., p. 150 (*New Science* ["NS"] 448).

21   Ibid., p. 22 (NS 34).

22   Ibid., pp. 112 and 116–17 (NS 367, 375–6 respectively).

23   Aristotle, *Aristotle in 23 Volumes*, vol. 23, trans. W.H. Fyfe (Cambridge, MA and London, 1932), *Poetics* (1457b 9–10).

24   Gaston Bachelard, *Poetics of Space* (Boston, MA, 1964), p. 74.

25   Ibid.

26   Ibid., p. 75.

27 Ibid., p. 8.

28 Certainly a house is not one of the four elements like fire mentioned by Illich previously. In the *Poetics of Space*, Bachelard meditates on how the character of spaces in a house contributes to our image of it. A similar comparison could be made using the image of a gutter in the metaphorical locution, "get your mind out of the gutter."

29 Juhani Pallasmaa, *The Embodied Image: Imagination and Imagery in Architecture* (Chichester, 2011), p. 67.

30 Calvino, "Lightness," *Six Memos for the Next Millennium*, pp. 3–29 and 8–10 in particular.

31 *David Leatherbarrow*, "Showing What Otherwise Hides Itself: On Architectural Representation," *Harvard Design Magazine* (Fall 1998): p. 50.

32 Marco Frascari, *Eleven Exercises in the Art of Architectural Drawing: Slow Food for the Architect's Imagination* (Abingdon, Oxon and New York, NY, 2011), p. 130.

33 Ibid., p. 6.

34 Ibid., p. 10.

35 Ibid., p. 90.

36 Jane Bennett, *Vibrant Matter: A Political Ecology of Things* (Durham, NC, 2010), pp. 13 and 18 respectively.

37 Ibid., p. 18.

38 László Moholy-Nagy, *The New Vision: Fundamentals of Bauhaus Design, Painting, Sculpture, and Architecture*, trans. D.M. Hoffmann (New York, NY, 2012), pp. 18–19.

39 Ibid., p. 60.

40 Adolf Loos, *Ins Leere Gesporchen* (Vienna, 1981), pp. 163–4 after David Leatherbarrow, *The Roots of Architectural Invention: Site, Enclosure, Materials* (Cambridge, England, 1993), p. 210.

PART ONE
The Material of Architectural Conceptions

1

Material Intuitions: Tracing Carlo Scarpa's Nose

Carolina Dayer

Tell me where and when I can see you once more. Or rather, see you for the first time![1]

Italo Calvino

Another late night at Carlo Scarpa's house and studio. The tired Guido Pietropoli, one of Scarpa's assistants, is tracing the plan drawing for the reconstruction and extension of the ex-convent of San Sebastiano to convert it for use by the Faculty of Literature and Philosophy of the University of Venice. Pietropoli is drawing

1.1 Carlo Scarpa. Reconstruction and extensions of the Convent of San Sebastiano, Faculty of Literature and Philosophy, University of Venice, Venice / Floor Plan 1974–78.

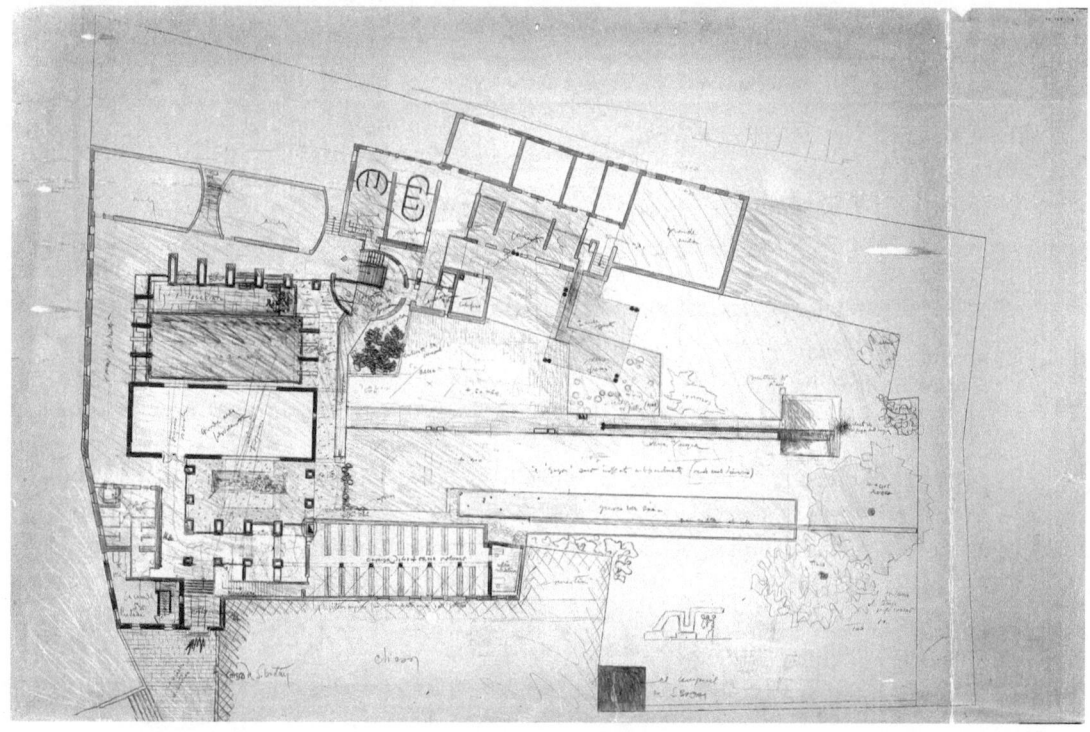

1.2   Carlo Scarpa.
Zoom-in detail.
Reconstruction
and extensions of
the Convent of San
Sebastiano, Faculty
of Literature
and Philosophy,
University of
Venice, Venice /
Floor Plan 1974–78.

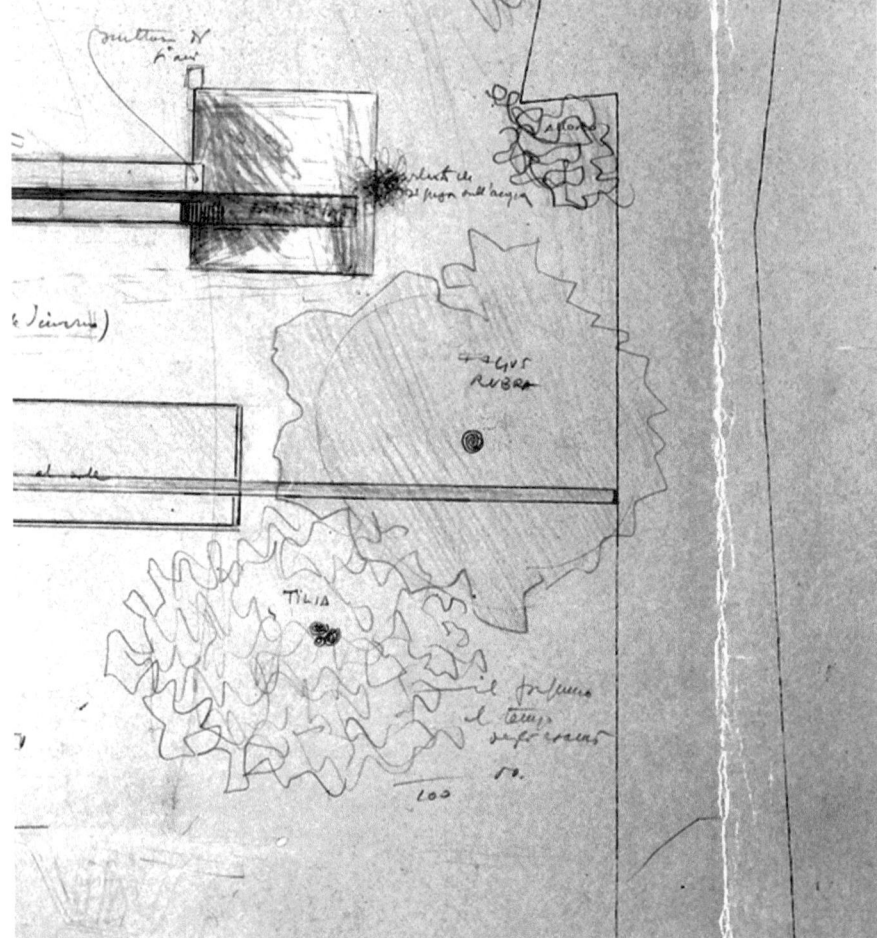

"tremblottant"—making free-hand ink lines with a thin nib by tracing over the constructed lines of the drawing below. The wavy pulse of his hand transfers through the nib into the paper, expounding slightly different qualities each time a line is traced. The entire drawing is made up of these thin lines. At first sight the line seems drawn with a ruler; however, with attention, subtle differences are discerned.[2] Each line faintly expresses particular movements of the hand, enlivening the drawing. The technique, according to Pietropoli, requires slowness and is a bit annoying; however, thanks to the slight variations, it imparts great character to the drawing.[3] While he is slowly and carefully drawing a long line in this way, the cigarette hanging from Pietropoli's mouth continues to burn until finally an ember of his cigarette falls onto the drawing. Before he can brush it away, the hot ember immediately burns through the tracing paper. Piercing through it, the ember leaves a perfectly round hole where it accidentally hits the paper. Pietropoli curses this loss of the night's work and is contemplating starting the drawing entirely over again when Scarpa sees it and exclaims, "… We will have a tree here!" and he draws a circle around the unexpected hole.[4] On the next plan for

San Sebastiano, the addition of the tree appears and is thoroughly integrated into the project (Figures 1.1 and 1.2).

These kinds of events regularly happen in architectural design, yet are never discussed in writings about architectural drawing. This omission in discussions of drawing, results in the incorrect assumption that architectural design flows from rational determination in the mind that is only afterwards recorded on a drawing. Perhaps it is merely a coincidence that the Saint Sebastian was often painted tied to a tree with numerous arrows piercing his body, though they did not kill him (Figure 1.3).[5] Similar to the arrows that did not kill the body of Saint Sebastian, so the ember piercing the plan did not end its life. As the personified figure of opportunity—*Occasio*—one must seize an opportunity when it presents itself. *Occasio*, who holds a knife, has long hair at the front so one can seize an opportunity as it confronts them, but is bald at the back of the head, because once an opportunity has passed, it usually does not return. Yet, the only way to be able to act quickly when the moment demands is by slow, thorough and careful preparation that proceeds the moment of fast action. Instead of errors that waste time in drawing, Scarpa's cunning mode of thought uses the opportunity to create something better. The slow making of the drawing and the fast burning of the paper staged a condition that many architects would consider a mistake, yet for the Venetian architect constituted a chance. A chance has an ambiguous meaning—it is an accident and it is an opportunity.[6] The accident must not be seen as motivated, that is, it cannot be seen as an economic justification to save the drawing. The accident must be seen as a crucial element already making the drawing.

The unexpected mark created by the ember became in Scarpa's hands not just a tree but a very specific one designated by Scarpa as a *fagus rubra*. *Fagus* is the Latin word for beech. Just like the ember consumed the paper, the word *fagus* was usually associated with edible fruit trees.[7] *Rubra* means red.[8] A *fag*, is also the British colloquialism for cigarette. While it is likely that Scarpa would know this word given his philological interests as well as his knowledge of the English language, we cannot give a precise reason as to why he chose the tree he chose. Perhaps Scarpa indicated red for the drawing's recreation of the spilling of the blood of Saint Sebastian, or perhaps for honoring the fiery color of the ember that generated the mark. In Scarpa's imagination the unexpected burn mark on the drawing was translated into a red beech, known for its blazing red foliage during the fall. Additionally, adjacent to the red/copper beech tree, the drawing showing yet one more tree: a *tilia*. Right next to it, Scarpa leaves a note that says, "… perfume for when students are in exam season."[9] The *tilia* or linden tree not only emits

1.3   Paolo Veronese. Virgin and Child Enthroned with Saints. San Sebastiano, Venice, 1564–65.

a delicious perfume when in blossoms but also holds in its leaves and flowers a delicious flavor recognized as carrying a calming effect when consumed in the form of tea. From the slow, wiggly construction of lines on a late night at the drawing table to the imagining of a place through olfactory vision, Scarpa's San Sebastiano plan demonstrates the possibility of chance encounters in the making of drawings.

INTUITION SNIFFS OUT POSSIBILITY

Beyond rational planning, when designing, the architect must use all aspects of bodily thinking since spaces are for complete inhabitation. This was understood by the renaissance architect Phillibert de l'Orme, who cautioned that, "the bad architect has little nose, for he does not have the intuition of good things."[10] The story of Scarpa's interpretation of the ember in his plan for the garden of San Sebastiano reveals an adoration and respect for the magical potential of mistakes and everyday life chances that occur in the work while drawing and thinking.[11] Intuition can be considered the architect's faculty of looking after the potential of every mark, action and accident that enters or surrounds the realm of the drawing. When exceptions are seen as fruitful distractions, the making of the drawing acts as a window that constantly re-frames the work into unknown and novel horizons. The rupture in the linearity of the work by a sudden mistake, like the ember burning the paper, allows the drawing to be concerned with two simultaneous realms: what is happening now, that is, the fire pricking through the sheet, and what will be happening later, the desire to make sense out of a senseless mark, that is, the translation of the hole into a tree.[12] As well, the meditation involved in the use of a technique such as *tremblottant* becomes a key point in opening up the architect's intuition to instances of chance encounters, where slowness allows one to rapidly apprehend possible accidents as opportunities for design.

The sense of smell has close associations with the notion of intuition. That intuition is often related with one's sense of smell, as when a friend advises, "follow your nose!" as a clue to ponder the exterior interiority and internal exteriority qualities of intuition and the potential for its cultivation. The etymology of the word intuition indeed expresses these two opposite actions, one moving inwards, the other moving outwards. While the Latin prefix *in* means "inside," the Latin word *tuērī* means "to look out" as well as "to look after, to watch and to protect."[13] Intuition is structured by an interiority and exteriority that simultaneously depend upon each other. Rarely thought as a source for knowledge that can be cultivated, intuition is commonly understood as a gifted internal vision or genius that is either present or not. However, the notion of the nose as a device for grasping the creative genius can be counterbalanced by looking at Scarpa's practices and the recognition of both common and exceptional everyday life events as potions for cultivating the imagination.

Paul Klee, an artist in whom Scarpa was quite interested, created a painting entitled *nach der Zeichnung* ("After the Drawing"), which has often been related to the introspection of creative genius (Figure 1.4).[14] Although Klee's drawing portrays the Swiss artist with an open nose, it has closed eyes, mouth and no ears. In the

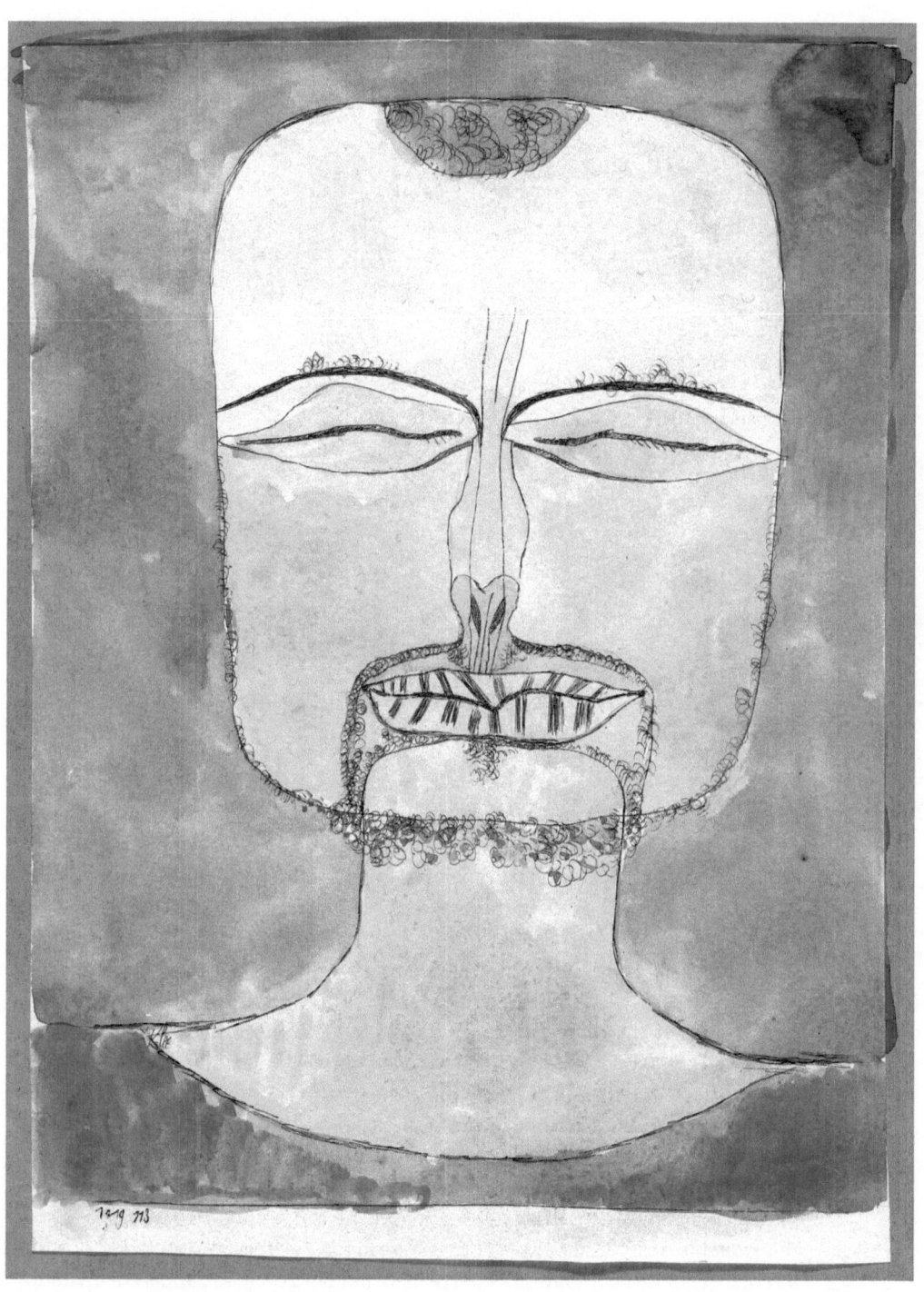

1.4   Paul Klee, *nach der Zeichnung* ("After the Drawing"), 19/75, 1919, 113.

1.5   Gianni Berengo Gardin. Venezia, Allestimento Biennale XXXIV: "Linee della ricerca: dall'informale alle nuove strutture," Padiglione Italia. Primo piano di Carlo Scarpa, 1968/06/25.

essay "Exact Experiments in the Realm of Art," Klee gives a defining role to intuition by noting, "We construct and construct … and yet intuition is still a good thing. A considerable amount can be done without it, but not all."[15] He considers the works of genius as something that cannot be accomplished without intuition, explaining that genius cannot be taught since it does not operate in the realm of rules. On the contrary, he states, it exists in the world of exceptions.[16] An exception is an action or phenomenon that is taken out or taken up from its regular course of being or behaving, that is, it means both to capture and to remove.[17] An exception discloses a break in the assumed sequence and order of things, and it can exhibit a completely unknown element or re-introduce something too evident that has remained unseen. According to Klee, genius operates in this realm, and for this reason it cannot be educated through a formal process. Towards the end of the essay, however, he notes that it is through a cultivation of exactitude that the foundations for a science of art can be laid.[18] Those foundations include the presence of unknown variables that are dependent on intuition. Yet, while examining his portrait, the completely introverted genius–artist, because he is alive, he is not as internalized as one may think, as his nose is still open, thus suggesting that the outside world and the internal life of the artist are still intertwined. Building on Klee's apparent invitation to look beyond an introverted notion of genius, this chapter argues that intuition must also be cultivated externally—smelled, perhaps—through an aperture to the world and to the materiality of everyday life.

If literally following one's nose is apprehending a new scent and desiring to know its source, metaphorically it is apprehending unexpected or common events of the everyday life in the making of the drawing and allowing the imagination to interact with them. Scarpa, who had a notorious elegant large nose, made sure to emphasize this virtuous feature of his face every time he depicted himself in drawings. (Figure 1.5). While he was a professor at the IUAV students believed that if they smelled Scarpa's favorite drink they would acquire the professor's genius.[19] Curiously, the drink, called *Underberg*, is a particularly scented digestive bitter produced in Germany with a secret recipe that includes aromatic herbs from 43 countries. Underberg is still advertised as a drink "to feel bright."[20] Like a magic potion, the bitter is characterized for being distributed in very small glass bottles wrapped in paper. The sense of smell, culturally and physically significant to human existence, is a mysterious sense that can trigger different associations within our brains concocting emotions, memories and desires that provoke the imagination in unpredictable ways.[21]

Reflecting upon these aspects of the sense of smell, parallels can be drawn for airing the notion of a cultivated intuition. In following our nose, for example, we only notice an odor when we smell an exception to the scents we have become accustomed to after a period of time. What we consider the norm, that is, "to smell," constitutes, for conscious perception, an exception within the constant act of smelling rooted in how we are in the world.[22] The exception only becomes apparent in the recognition of the notes of a scent. However, what may be an exception at one instant can constitute the norm an hour later once one has become used to the notes of such scent. A clear example of this is when we enter a house that is not ours and we smell the house's aromas, yet the ones living in it have ceased to smell them.[23] In this way, following Klee's observations from above, exceptions are determined by how we experience them and not by what they are as an absolute condition. Both strong and subtle odors have the potency of emerging into the extraordinary and disappearing into the ordinary. Our nose acts as an essential probe in the discovery of the unexpected, as well as in the conscious wondering of the norm. Our nose can also be educated, even if it cannot be formalized into recipes. Practices like wine tasting recommend the smelling of spices, flowers, woods, fruits and other materials to prepare the nose for the moments of wine tasting. Knowing the odors is a crucial aspect for recognizing each of the many flavors one sip of wine can have.

When making architectural drawings, the architect constantly experiences "exceptions" that distract, interact with, and provoke the imagination at minimal and significant levels.[24] From the wind entering into the room while drawing to a heated argument with the client, the architect's work is always "interrupted" by something. A cultivated intuition is the capacity of the imagination to find potency in instants of interruption or suspension that detour the expected and become active in the realm of the creative. What is more, these moments of interruption are crucial for thinking, and we rely on our nose both metaphorically and physically to engage creatively with them. When the particular apparition and recognition of common events are handled as exceptions or exceptional events are treated as the norm—when the nose is awakened—intuition finds place for imaginative actions. The instant when something appears or something is recognized as a chance for an imaginative action is a quality that architects can cultivate by being willing to see what reality hides.[25] The possibility that there is always something hiding behind reality re-positions intuition as an active trait of the imagination and not a passive gift of genius.

DISTRACTED ATTENTION

In his book *Intuition of the Instant* from 1932, Gaston Bachelard, interested in developing a theory of time through the recognition of the "instant," argues that instants are the only reality of time.[26] Once smelled, an odor becomes immediately diminished over a short period of time. The structure of the instant embraces a snap of time that obscures and denies the horizontal linearity of time, concentrating on the verticality of the here and now.[27] By looking at the fertility of an instant, a special ability of the nose, a common or accidental event holds the potential to become exceptional while an extraordinary event upholds the possibility to be discreetly veiled within the norm. Bachelard describes an instant as a series of simultaneities that meet and join, flipping the understanding of time from horizontal to vertical.[28] He describes the instant as ontologically poetic and "necessarily complex: it moves, it proves—it invites, it consoles—it is astonishing and familiar. It is essentially a harmonic relation between two opposites."[29]

Because we can experience and exist within all these simultaneities, Bachelard's theory of the instant is understood as vertical and not linear. Like the medium of smell and air, the instant is constituted by depth and height; it must be inhabited.[30] The meaning of the word "instant" is that of being present, thus if we place intuition where the imagination acts within the reality of an instant, the relationship between *what* is happening and *how* I am experiencing it stands essential. Intuition, just like the nose, is seduced as long as things or scents manifest themselves within an instant.[31] But how can one cultivate a readiness that will allow intuiting something hidden within an instant? Bachelard shows another important concept playing a role within the structure of the instant. That is the role of attention. He claims that, if we consider attention directed to life and not to private thoughts, attention is always born of coincidence and not of planning. Coincidence is what brings novelty and what he assures is necessary for the mind to focus.[32] Contrary to the assumption that attention is a planned and controlled intention, the French philosopher offers a fresh thought where the "clarity of motive and the joy of acting suddenly converge."[33]

In the case of San Sebastiano, Scarpa focuses on a commonly recognized nasal expression "follow your nose" as when one suspects that an unexpected mark must be hiding something. In following his nose, Scarpa saw that a hole in the paper was hiding a possible tree in the garden. Both smell and vision, in a synesthetic marriage, allow the architect to focus on the potency of accidents as the work is made and it is simultaneously making itself. In Bachelard's notes on attention there is a value given to will and decision-making that is conscious within the instant.[34]

While unexpected appearances are aspects of the instant, they are only revelatory if there is a conscious apprehension and re-positioning of their qualities. Our everyday life reality is soaked with these happenings. Exceptions, errors and other unusual instances of the everyday life can intensify the potential of the cultivated intuition to act. Nevertheless, very common and usual events can also trigger creative actions. A "distracted attention" is desirable when one wants to be aware of what reality hides and to be distracted from what it reveals.

One example illustrating the actions of intuition within the fabric of everyday life events that are not exceptions is told by Francesco Zanon, one of the smiths who worked with Scarpa at the Brion Cemetery. Francesco tells the story of a particular detail that took the architect more than two months to design. The detail consisted in the installation of a tensed metal rod cable that had to span through the entire width of the garden. Many *in situ* tests were performed, but the tensed cable kept snapping, especially when kids on Sunday would play and jump on it. Francesco recalls the day when Scarpa called him and said: "Francesco: I have solved it!" Francesco later discovered that the idea had come to the architect by observing the large metal springs commonly used in trains, Scarpa's normal mode of travel between places. The story exposes a form of creative thinking intimately tied to his everyday life. Francesco curiously describes this instance of discovery as the intuition of the architect.[35] In an earlier interview, Francesco's brother, Paolo Zanon, expresses that Scarpa always wanted to be informed regarding new materials or how to treat existing ones in new ways. In the interview he recounts that most of the time, "Scarpa was not satisfied with materials commonly found on the market, and sometimes he had the need to design handcrafted details that went from the handles to the hinges."[36] Always using the events of everyday life as a point of departure, the Venetian architect was dedicated to making magnificent unexpected connections between the work and quotidian objects of the everyday.

Through making connections, Scarpa was also able to care for moments in the life of a drawing that could reveal something unknown. He believed in capturing accidental marks that could allow him to imagine new presences, a kind of knowing without knowing, as in his plan for San Sebastiano. Another example of the architect's cultivated intuition is evident in the design and construction of the Gipsoteca Canoviana, a museum to host Antonio Canova's sculptures at Possagno. In this project Scarpa encountered a design mistake while one of the windows was under construction. In a lecture he gave to students at the IUAV in 1976, he tells the story of a skylight window in the shape of a cube he designed in order to bring the sky into the room. While designing the architectural detail it never occurred to him that the reflection between the glass panes would perform another reality, taking away the clarity existing between the transparent solidity of the glass and the various airy hues of the sky.[37] Scarpa after this unnoticed condition concludes:

> These are the errors that allow one to incur into thinking, acting, and making, and when they happen you need to have a double mind, a triple one, the mind of the thief who speculates the robbery of a bank and needs to have that which I call wit, an attentive tension to understand all what is happening and all that will happen.[38]

The attentive tension suggests the presence of an awareness and readiness to act and imagine. When working with materials, one must consider how they are and interact in the world under many conditions. Contrary to a focused attention to one thing, the attentive tension is a state of approaching the work and the world that allows the architect to imagine with the unexpected, in this case the reflectivity of the glass was an ingredient to be considered when designing the detail.

FACTS, ACTS, MARKS

Architects' imagination, contrary to the common acceptance of being internal, highly depends on external events and it relishes considering the unpredictable as a potent tool that makes the work as the same time that allows it to be made. Bachelard writes, "facts, besides being facts, are acts. And acts, however unfinished or unsuccessful, must necessarily begin in the absolute of a birth."[39] By apprehending and protecting the unpredictable "mistake," as in the burning of a cigarette ember in the project of San Sebastiano, for example, Scarpa celebrates its birth and incorporates it into the life of the design. This is not a life for the idle visitor or unaware architect, but for the active designer, who realizes in the design an extended potentiality through the inclusion of unexpected events. In addition, the "mistake" aids the arrival of other actions, and it invites the possibility of odoriferous presences, such as the linden tree especially chosen for the students inhabiting the garden, a detail that would be inexplicable without the fortuitous intrusion of the ember into the drawing. By looking after his mistake, Scarpa takes care of the design and its future inhabitants. In Bachelard's words, "novelty is needed for thought to intervene; novelty is also needed for consciousness to affirm itself and for life to progress."[40] The unexpected new mark in the drawing manifests an exception in the life of the project that detours the imagination of the architect, thus exposing it to new opportunities for the design.

For Marco Frascari, who ruminated on the virtues of architectural drawing, "every mark matters."[41] This is certainly true of Scarpa, who willingly considered the actions of everyday life as a part of the drawing itself. When Frascari expresses, "creativity is an experience, not an abstract idea that the mind and body incessantly analyzes," he suggests that creativity works within how one experiences the directness of ordinary encounters.[42] The architectural thinking that desires a "straight" line, like the *tremblottant*, a line in fact full of imperfections and non-straight lines, is the same kind of architectural thinking that is required to imagine that, from an injury in the paper, the life of a tree can participate in the design. In other words, an intuitive architect is the one who attends carefully to all the marks that will mark the life of the building. Likewise, an intuitive architect is the one who believes that a straight line is never "just a straight line" and a hole in a drawing is never "just a hole in a drawing." There is always something more, that is, a life in marks that exists beyond architect's intention. In the introduction of his book *Air and Dreams*, Bachelard states that, "if the image that is present does not make us think of one that is absent, if an image does not determine an abundance—an explosion—of unusual images, then there is no imagination."[43] The mark of the accident acts as an image that re-frames itself within the drawing and the architect discovers it as already being a different entity, in this case, a tree in the garden.

In architecture, the practice of "following one's nose" also has an ethical dimension, that of caring for and protecting the design. Due to the variable conditions of an architectural project, constant shifts, revisions and alterations partake in the life of the edifice. The precise and careful construction of drawings has the potential to lead to very imprecise marks that provoke the imagination. When these marks are treated as gifts and not mistakes, the imagination finds places for them and they find a place within the architect's imagination.

Scarpa gives a curious definition of intuition himself. He understands undefined marks in a drawing as "surprises of the imaginative plot of signs that come, sometimes from ideas, if one has the intuition to understand the everyday life, and this type of intuition is equivalent to that which allows one to construct many thoughts while looking at the hair of a woman."[44] Scarpa's definition of intuition reveals his quest for an active imagination that composes possibilities from disruptions of normal events, or simply for re-framing those normal events into new ones. Attention to everyday life events and phenomena, as was demonstrated previously, can offer new encounters and trigger novel or originative conceptions not just within the mind of the architect,[45] but within the material reality of the work itself. Scarpa finds a materiality outside his mind, in the drawing itself, something he then realizes the design needed.

In this way, Scarpa shows that the primacy of the instant, as a place able to create turbulence within the linearity of architectural drawings, exposes the necessary presence of everyday life events participating with the imagination. Initial marks in an architectural drawing have the potential to be understood differently through unexpected happenings occurring during the act of drawing. The architect with a cultivated intuition is able to re-frame and re-think previous calculated design decisions into new joyful occurrences. This is very different from attempting to explain the current state of a drawing through its past conditions, where the final design is understood as a total calculated process.[46] In such a case, for example, a mistake would only remain a mistake and not a poetic reconstruction. Re-positioning the works of the imagination in the field of the drawing itself and not inside the mind allows the drawing to take on a life on its own where the architect's acts, planned and unplanned, can interact and emerge into unexpected designs.

THE WORLD IS THE NOSE

In the short story by Italo Calvino, *The Name, The Nose*, the Italian writer begins by foreseeing the man of the future as noseless, hypothesizing a world where perfumes could not be recognized any longer. The story contains three stories of desire, search and loss. The three stories portray the sense of smell as a form of knowledge, so much so that one of the characters expresses: "We understood whatever there was to understand through our noses rather than through our eyes ... everything is first perceived by the nose, everything is within the nose, the world is the nose."[47] This character in the story is a primate that, just like the other characters that are human, is also in search for that one smell that will fulfill a higher desire. In the three intertwined stories, the characters, through the material particularities of scent, move in the world confident of their skilled noses.

With the same care for details, intensities and properties of the materiality of everyday life, Carlo Scarpa cultivated a readiness and awareness towards the unexpected ventures that being in the world could bring to his imagination. The intuition of the instant, as a place able to interrupt the linearity of architectural drawings, exposes the necessary presence of everyday life events participating with the imagination. The previous marks in an architectural drawing may be understood through its present epiphanies occurring during the act of drawing, thus allowing the architect to re-frame previous calculated design decisions.

A cultivated intuition is one that protects these marks or instances and receives them as poignant chances for the imagination. The novel odor of burnt paper as well as a little hole in the "wrong" place certainly invited the reorientation of the imagination to act upon the new conditions of the drawing. The short story of Scarpa's work on the garden of San Sebastiano demonstrates a mode of working and thinking founded on the inclusion of all the moments, events and acts happening as the drawing is constructed. This relationship with the work provides an essential ingredient for the architect's intuition. Because intuition lingers in regions where time, material and reason collapse, becoming attentive to "mistakes" presents new possible detours that may offer a more affable destination.

BIBLIOGRAPHY

Andrews, E. A., Charles Short, Charlton Thomas Lewis, and William Freund. *A Latin Dictionary Founded on Andrews' Edition of Freund's Latin Dictionary*, rev., enl., and in great part rewritten by Charlton T. Lewis and Charles Short (Oxford: Clarendon Press, 1962).

Bachelard, Gaston, *Air and Dreams: An Essay on the Imagination of Movement*, trans. E.R. Farrell and C.F. Farrell (Dallas, TX: The Dallas Institute Publications, 1988).

Bachelard, Gaston, *Intuition of the Instant*, trans. E. Rizo-Patron (Evanston, IL: Northwestern University Press, 2013).

Calvino, Italo, *Under the Jaguar Sun*, trans. W. Weaver (Orlando, FL: Harcourt Inc., 1983).

Certeau, Michel de, *The Practice of Everyday Life*, trans. Steven Rendall (Berkeley, CA: University of California Press, 1984).

Chupin, Jean-Pierre, "Hermes' Laugh: Philibert de l'Orme's Imagery as a Case of Analogical Edification," in A. Perez-Gomez and S. Parcell (eds), *Chora 2: Intervals in the Philosophy of Architecture. Vol. 2* (Montreal: McGill-Queen's University Press, 1996).

Corbin, Alain, *The Foul and the Fragrant: Odor and the French Social Imagination* (Boston, MA: Harvard University Press, 1982).

Dugan, Holly, *The Ephemeral History of Perfume: Scent and Sense in Early Modern England* (Baltimore, MD: The Johns Hopkins University Press, 2011).

Emmons, Paul, "The Place of Odor in Modern Aerial Urbanism," *Journal of Architecture (RIBA Journal of Cambridge University Press)*, Vol. 19/2, *Special Issue: Urban Air* (2014).

Frascari, Marco, *Eleven Exercises in the Art of Architectural Drawing* (London: Routledge, 2011).

Giesche, Frederick E., *Technical Drawing* (New York, NY: The Macmillan Company, 1940).

Giordano, Sandro, *Il mestiere di Carlo Scarpa: collaboratori, artigiani e committenti*, relatori: F. Dal Co, G. Mazzariol. Doctorate Thesis (Venezia: IUAV: Istituto universitario di architettura di Venezia, 1984).

Hairston, Julia L. and Walter Stephens (eds), *The Double Life of St. Sebastian in Renaissance Art: The Body in Early Modern Italy* (Baltimore, MD: The Johns Hopkins University Press, 2010).

Klee, Paul, *Notebooks, Volume 1, The Thinking Eye*, trans. R. Manheim (London: Lund Humphries, 1961).

Linzey, Mike, "On the Secondness of Architectural Intuition," *Journal of Architectural Education*, Vol. 55/1 (September 2001): pp. 43–50.

Mauss, Marcel, *A General Theory of Magic*, trans. R. Brain (London: Routledge Classics, 2001).

Merleau-Ponty, Maurice, *Phenomenology of Perception*, trans. C. Smith (London: Routledge & Kegan Paul, 1962).

Ohloff, Günther, *Scent and Fragrances: The Fascination of Odors and their Chemical Perspectives* (Berlin: Springer-Verlag, 1994).

Rosand, David, *Drawing Acts: Studies in Graphic Expression and Representation* (New York, NY: Cambridge University Press, 2002).

Semi, Franca, *A Lezione con Carlo Scarpa* (Venice: Cicero, 2010).

Teuber, Dirk, "Intuition und Genie: Aspekte des Transzendenten bei Paul Klee," in Paul Klee, Manfred Fath, and Hans-Jürgen Buderer (eds), *Paul Klee, Konstruction, Intuition* (Stuttgart: G. Hatje, 1990).

Werckmeister, O.K., *The Making of Paul Klee's Career 1914–1920* (Chicago, IL and London: University of Chicago Press, 1989).

NOTES

1   Italo Calvino, *Under the Jaguar Sun* (Orlando, 1988), p. 78.

2   In a United States 1940 manual for technical drawing, this type of line is called a "straight line for technical sketching." While the instructions of the manual explain the making of the line with the use of pencils, the procedure nearly mirrors the one drawn with ink. According to this manual, the making of the line must be informed by "freedom, snap and confidence," and it advocates for paying attention to the tip of the pencil or nib while making the line instead of focusing on the entire line. Giesche, Frederick E., *Technical Drawing* (New York: 1940), pp. 517–20.

3   Interview by author to Guido Pietropoli, July 2013. Pietropoli learned the drawing technique in the Atelier of Guillermo de la Fuente in Venice while working on the Venice Hospital project for Le Corbusier. When Scarpa heard of the technique, he asked Pietropoli to use it in the drawings for San Sebastiano.

4   I greatly thank Guido Pietropoli for telling me this story while we were observing Carlo Scarpa drawings during the exhibition: Progetti veneziani di Carlo Scarpa: le università at Archivio Progetti Iuav curated by a cura Serena Maffioletti and Archivio Progetti— SBD with Leonardo Monaco and Mara Micol Reina in Venice on July, 2013.

5   *The Double Life of St. Sebastian in Renaissance Art: The Body in Early Modern Italy*, ed. Julia L. Hairston and Walter Stephens (Baltimore, 2010).

6   The etymology of the word chance has its roots in the Latin *cadentia*: falling. In addition to the two meanings, accident and opportunity, the word can also mean a space in time. *Oxford English Dictionary*, Online Dictionary, Oxford University Press, 2014. Available at: www.oed.com [accessed 25 August, 2014].

7   Andrews, E.A., Charles Short, Charlton Thomas Lewis, and William Freund. *A Latin Dictionary Founded on Andrews' Edition of Freund's Latin Dictionary*, rev., enl., and in great part rewritten by Charlton T. Lewis and Charles Short (Oxford, 1962).

8   I would like to thank the landscape architect Nathan Heavers for his insights about the qualities of this tree.

9   The original Italian note reads, "*Il profumo al tempi degli esami.*"

10   Jean-Pierre Chupin, "Hermes' Laugh: Philibert de l'Orme's Imagery as a Case of Analogical Edification," in *Chora 2: Intervals in the Philosophy of Architecture. Vol. 2*, ed. A. Perez-Gomez and S. Parcell (Montreal, 1996), p. 44, translating Philibert de l'Orme, *Premier tome*, 28v.

11   Magic in the sense that something external can alter or change the natural course of action of something else. As Marcel Mauss has expressed, "magic is the art of changing." Marcel Mauss, *A General Theory of Magic*, trans. Robert Brain (London, 2001), p. 76.

12   Mike Linzey, "On the Secondness of Architectural Intuition," *Journal of Architectural Education*, Vol. 55/1 (September 2001), p. 43. He also expresses the ambiguity of the notion of intuition as "a kind of awareness of relations that is intermediate between sensory perception and conceptual thought …." Additionally, "The intuition of relations can be an object of wondrous intent and curious inquiry that sometimes only dreaming and idle mediation reveals …," p. 45.

13   The Oxford English Dictionary defines the term as: insight, internal vision, quick apprehension of something that is contemplated in the mind, etc. *Oxford English Dictionary*. Available at: www.oed.com [accessed 25 August, 2014].

14   O.K. Werckmeiser, *The Making of Paul Klee's Career 1914–1920* (Chicago, IL and London, 1989), pp. 205–7.

15   While Paul Klee states in his essay that genius/intuition cannot be taught, there is yet a suggestion that, by being in the world, experiencing nature, the things that are "unexplainable" participate and are part of the creative actions of the artist. Paul Klee, *Notebooks: Volume 1, The Thinking Eye*, trans. Ralph Manheim (London, 1961), pp. 69–70.

16   I thank Holger Gladys for helping me translate an article that expanded my understanding of Klee's notion of intuition by Dirk Teuber: "Intuition und Genie: Aspekte des Transzendenten Bei Paul Klee," in Paul Klee, Manfred Fath, and Hans-Jürgen Buderer (eds), *Paul Klee, Konstruction, Intuition* (Stuttgart, 1990), pp. 33–43.

17   Exception: French *excepte-r*, < Latin *except-* participial stem of *excipĕre* to take out, < *ex-* out + *capĕre* to take. The formally equivalent Latin *exceptāre* had only the sense "to catch, take up." *Oxford English Dictionary*, Online Dictionary, Oxford University Press, 2014. Available at: www.oed.com [accessed 25 August, 2014].

18   The "exact" in the essay makes reference to what he calls an "exact research," that is, mathematics, physics, mechanics, history, nature, but also things like "learning to look down on formalism and to avoid taking over finished products, or leaning to organize movement to logical relations." Klee, *Notebooks: Volume 1, The Thinking Eye*, p. 69.

19   I thank Guido Pietropoli for telling me this beautiful story in my quest to learn about Carlo Scarpa's nose.

20   Underberg website. Available at: http://www.underberg.com/en/home.html [accessed 2 November, 2014].

21   Holly Dugan states the reports of the Nobel Prize in Physiology or Medicine of 2004 as key for describing the complexity of smell for research and science as well as for cultural studies. Holly Dugan, *The Ephemeral History of Perfume: Scent and Sense in Early Modern England* (Baltimore, 2011), pp. 1–23.

22   Maurice Merleau-Ponty, *Phenomenology of Perception*, trans. C. Smith (London, 1962), pp. 235–46.

23   Günther Ohloff, *Scent and Fragrances: The Fascination of Odors and their Chemical Perspectives* (Berlin, 1994), pp. 57–8.

24   Donald Kunze expresses: "Thus, it is important for architecture theory to pay its closest attention to cases where representation breaks down, for it is precisely at such points that invisibility becomes critical." Donald Kunze, "Metalepsis of the Site of Exception" (unpublished essay). Available at: http://art3idea.psu.edu/ [accessed 11 February, 2014].

25   The notion of something hiding within reality constitutes a larger research project on Magic Realism, the work of Carlo Scarpa and notion of reality explored by the Neapolitan philosopher Giambattista Vico. I thank Donald Kunze for his immense help and guidance on these matters.

26   Gaston Bachelard, "Poetic Instant and Metaphysical Instant," in *Intuition of the Instant*, trans. E. Rizo-Patron (Evanston, 2013), p. 6.

27   Ibid., Appendix: "Poetic Instant and Metaphysical Instant by Gaston Bachelard," pp. 19, 58.

28   Ibid., p. 59.

29   Ibid.

30   Dr. Paul Emmons has written, "Since air is the material medium through which humanity swims, Le Corbusier emphasizes its importance with statements such as 'To live! (to breathe).'" Paul Emmons, "The Place of Odor in Modern Aerial Urbanism," *Journal of Architecture (RIBA Journal of Cambridge University Press)*, Vol. 19/2, *Special Issue: Urban Air* (2014): p. 208.

31   Instant: Etymology: French *instant* (14th cent. in Hatzfeld & Darmesteter) assiduous; at hand, imminent; < Latin *instânt-em*, present participle of *instâre* to be present, to be at hand; to urge, press upon; to apply oneself to; < *in-* + *stâre* to stand. *Oxford English Dictionary*, Online Dictionary, Oxford University Press, 2014. Available at: www.oed.com [accessed 25 August, 2014].

32   Bachelard, *Intuition of the Instant*, p. 21.

33   Ibid., p. 20.

34   Bachelard, "Poetic Instant and Metaphysical Instant," p. 21.

35   Francesco Zanon monologue in the documentary film *Memoriae Causa* by Riccardo De Cal, produced by Fondazione Benetton Iniziative Culturali, SOSS Film, in Italy (2007).

36   Interview to Paolo Francesco in Sandro Giordano, *Il mestiere di Carlo Scarpa: collaboratori, artigiani e committenti*, relatori: F. Dal Co, G. Mazzariol; [laureando]: Sandro Giordano, – 1 v.; 30 cm Doctarate Thesis (IUAV: Istituto universitario di

architettura di Venezia, Corso di laurea in architettura, Anno accademico 1983/1984, Sessione estiva, 1984), p. 98. Translation by author.

37   Franca Semi, *A Lezione con Carlo Scarpa* (Venice, 2010), pp. 191–2.

38   "Sono errori in cui si incorre nel pensare, nell'agire, nel fare, e quindi bisogna avere la mente doppia, la mente tripla, la mente del ladro, da uomo che specula, da uomo che vorrebbe rubare in una banca, e bisogna avere quel che io chiamo arguzia, una tensione attenta per poter capire tutto quel che succede e quello che succederà." Ibid., p. 192.

39   Bachelard, *Intuition of the Instant*, p. 13.

40   Ibid., p. 15.

41   Marco Frascari, *Eleven Exercises in the Art of Architectural Drawing* (London, 2011), p. 27.

42   Ibid.

43   Gaston Bachelard, *Air and Dreams: An Essay on the Imagination of Movement*, trans. E.R. Farrell and C.F. Farrell (Dallas, 1988), p. 1.

44   "e dalle sorprese del'immaginifica trama di segni che vengono, qualche volta vengono delle idee, se uno ha intuit da capire le cose della vita, l'intuizione che permette di far tanti pensieri anche guardando la capigliatura di una donna'." Semi, *A Lezione con Carlo Scarpa*, p. 225. Translation by author.

45   Michel de Certeau, *The Practice of Everyday Life*, trans. S. Rendall (Berkeley, 1984), pp. 125–30.

46   "We must therefore attempt to understand the past through the present, which is different from striving ceaselessly to explain the present through the past." Bachelard. *Intuition of the Instant*, p. 11.

47   Calvino, *Under the Jaguar Sun*, p. 71.

2

Extracting Desire: Michelangelo and the *forza di levare* as an Architectural Premise

Jonathan Foote

> *it was necessary for him to find another block of marble, so that he could*
> *continue using his chisel every day …* [1]
>
> Giorgio Vasari, referencing Michelangelo

Michelangelo Buonarroti, who died in Rome at age 88 after having carved marble only a few days prior, offered barely a handful of words during his long life about what he actually thought about the creative process. Among the paucity of written reflections was his response to the *paragone* debate, a topic of courtly discussion that centered on the relative status of painting and sculpture among an emerging class of artist–scholars.[2] In framing his retort Michelangelo famously distinguished between painting and sculpture based on their fundamental differences in technique. He wrote, "By sculpture I understand that which is done *per forza di levare* [by taking away]; that which is done by addition is similar to painting."[3] Following this characterization of the two arts, Michelangelo ultimately argued for the superiority of sculpture over painting based on the physical arduousness required by the sculptor working in obdurate material.[4] With this, he makes an overt connection between the difficulty of working with material and the emergence of conceptual thought, a theme that appears in one of his best-known verses:

> *Just as, per levar [by taking away] lady, one puts*
> *into hard and alpine stone*
> *a figure that's alive*
> *and that grows larger wherever the stone decreases,*
> *so too are any good deeds of the soul that still trembles*
> *concealed by the excess mass of its own flesh,*
> *which forms a husk that's coarse and crude and hard.*
> *You alone can still levarne [take them out]*
> *from within my other shell,*
> *for I haven't the will or forza [strength] within myself.* [5]

The technical imperative of removing excess material as a path toward poetic and moral transcendence has been well documented through analyses across

Michelangelo's *oeuvre*, perhaps best recorded in the *Boboli Slaves* left unfinished in Florence from 1534. In the above sonnet, Michelangelo asserts the fundamental power of material to shape the sculptor as much as the sculptor shapes the stone. What begins as an assertion of the passive material that receives the injection of an active, male spirit, results instead in the opposite, where it is the lady herself (as matrix) who removes the excess mass from the artist's soul. Whether it is the artist working the material, or the material the artist, the principal instigator for Michelangelo is the process of removal, or the *forza di levare*.

As an investigation into the material imagination, the notion that a deeply corporal process of removing excess material may provoke a conceptual imperative holds the potential to invigorate material reveries. It was within this context that Paul Valéry held up Michelangelo's sculptural practices as an exemplar for "improvisation on a higher scale," writing that, "between intention and means, between the fundamental conception and the act of engendering form there is no longer a barrier. Between the artist's mind and his material an inner conjunction has taken place ...."[6] The critical reciprocation between a solid, material matrix and the act of conception, well explored within Michelangelo's sculptural works, has yet to be tested against his architectural practices.[7] The historian Charles De Tolnay touched on this dialectic when he wrote of the Medici Chapel that, "the tomb architecture is a mass with emphasized projections and recesses, animated by its inner forces, and the figures seem incarnations of the vital energies potentially present in this mass itself."[8] Beyond such a revelation of form, however, one wonders if the *forza di levare* may also be present in the means by which he practiced and represented architecture. How he imagined the material actions carried out by others in the process of edification may also be seen as a vitalizing agent.

MODANI: CUTTING STONE/CUTTING PAPER

Among Michelangelo's rather extensive archive of architectural drawings, one particular set, known as *modani*, offers initial clues into how the *forza di levare* may be traced in his architectural practices. Drawn as paper templates, *modani* were 1:1 profile drawings of building cornices, architraves, and column bases, subsequently cut with scissors along their profile and passed from the drawing board to the stone masons on the building site. Once there, they would be traced directly onto a rigid support medium, such as wood or metal, after which they would act as 1:1 tracing devices for carving stone details. The most robust surviving set of *modani* remaining from the sixteenth century stems from Michelangelo's own hand for his three architectural projects at San Lorenzo in Florence including the never-completed façade, the Medici Chapel, and the Laurentian Library.[9] These template drawings offer an enticing window into his working methods, since they were both conceptual tools as well as devices having direct, practical use on the construction site. As sheets of paper prepared specifically for the stone carvers, their making may offer insight into how Michelangelo imagined cutting the stone through the hands of others.

Several key points emerge when examining the *modani* within the context of the *forza di levare*. First, although *modani* were drawn using the conventional tools of architectural drawing, such as ink or charcoal, they assumed their identity as *modani* after being seized upon by scissors and cut along a profile line. Second, as 1:1 drawings, they could be traced directly on a squared piece of stone, and they did not rely on any scalar translation to be useful as a construction implement. Absent of being able to carve the stone himself, paper acted as a surrogate for Michelangelo to facilitate his subtractive method in imagining the stone detail. With this in mind, the well-documented figural character of his profiles may be seen not only through experiments in form, but also as a feat of imagination actively engaged during the material transformation. While making *modani*, Michelangelo relied on the same conceptual imperative and cutting practices to ultimately conceive the details. In this way, the potential of paper to act as a substitute material in place of stone was exploited to leverage the immediacy of his imagination onto the distant construction site.

Interestingly, Michelangelo's templates present no evidence of being constructed with compass and rule, a curiosity that differs drastically from his contemporaries, such as Antonio da Sangallo il Giovane and Bartolomeo Ammannati. The lack of marks, compass pricks, and dry point lines thus have generally led scholars to conclude that they were drawn free hand and without the aid of any constructive device. In fact his *modani* were a result of a sequence of cuttings and tracings, where the artist cut and then employed the templates as tracing devices to generate new *modani*. During the tracing process, Michelangelo stretched or slid the template, flipped it, or even assembled entirely new profiles using edge tracings from multiple templates. These procedures would have obviously eschewed residue from the compass and rule, as each *modano* was a descendant of both parent and off-spring template drawings that could be traced. From this he could use the paper to freely stretch and re-assemble the classical vocabulary beyond the bounds of measured conventions, while still being guided by constructive methods requiring sequence and tooling.

Understanding the full significance of Michelangelo's paper *modani* as devices to imagine specific working methods for assailing a stone block becomes an important point of comparison. Fortunately, several contemporary sources record his highly admired yet somewhat idiosyncratic technique. Unlike many sculptors who carved the block from multiple sides, moving around the block and slowly removing material, Michelangelo assailed a block from the front only, as if it were a bas-relief, revealing the figure from front to back. To this end, Benvenuto Cellini, in his *Trattato della scultura* of 1568, wrote, "the best method [for carving] I have ever seen was the one that Michelangelo used: after you have drawn your principal view, you begin to sculpt it with the virtue of the chisels, as if you were to make a figure in a half relief, and thus, little by little, it is sculpted."[10] And Vasari, in his *Life of Michelangelo*, described how Michelangelo sculpted the *Boboli Slaves* by, "first sculpting the parts of highest relief, and then little by little revealing the lowest parts."[11] He continues to describe how the artist would submerge a scaled model in water and then slowly raise it, acting as a guide by always delineating a frontal relief.

2.1   Cornice stone, carving in progress, Palladio Museum, Vicenza.

Interestingly, the frontal method described by Vasari and Cellini parallels how a block of squared stone is cut to produce a cornice section. Following the initial tracing of the *modani* on the vertical short side, the mason proceeds away from himself, lengthwise along the block, as a finished profile slowly emerges, little by little. As may be seen in a modern demonstration of this at the Palladio Museum in Vicenza, the relation of the finished profile to the stone matrix stands in remarkable similitude to what may be observed in the *Boboli Slaves*. One can imagine that the emergent cornice detail appears to be "escaping" the confines of the stony matrix.

In addition to articulating Michelangelo's frontal method, Cellini also asserts that one should first trace the principle view onto the block before cutting. Presumably, this initial drawing would provide the guiding lineaments for the sculptor to begin carving. A hidden figure enveloped by stone is captured in sketches by Michelangelo for his *Dei Fluviali* [River Gods], where a ghosted, reclining figure floats within the marmoreal confines of a large block. Upon close inspection, the sketches are found to be of the same figure in two different views, suggesting that it was a preparatory drawing exploring this technique. Peering inside one of these blocks and anticipating the removal of excess material recalls the opening quatrain of one of Michelangelo's most known sonnets, where he reflects on the reciprocity between the block of marble and the conception, or *concetto*:

> Not even the best of artists has any concetto [conception]
> that a single block of marble does not contain
> within its excess, and that is only attained
> by the hand that obeys the intellect.[12]

Any tracing on a stone block would likely be chipped away rather quickly; indeed, such lines act as a method for projecting deep into the block rather than as a step-by-step guide for the chisel. Just as Michelangelo first made a drawing on the stone to guide the initial chisel strikes, one could argue, so did he trace each *modano* as an invitation for the scissors. Either in stone or paper, the artist follows the sequence: trace, then cut.

The potential of the cutting tool to act as a creative instrument beyond the initial tracings seems present in paper as well as stone.[13] What may be observed on several *modani*, for instance, is a deviation between the traced or marked line and the line followed by the cutting tool. In such cases, Michelangelo used the *forbici* [scissors] on paper as a chisel on stone, where movements produced definitive cleaves and cuts that did not follow a pre-determined line or drawing mark. On a *modano* destined for a cornice detail in the Laurentian Library, for example, the

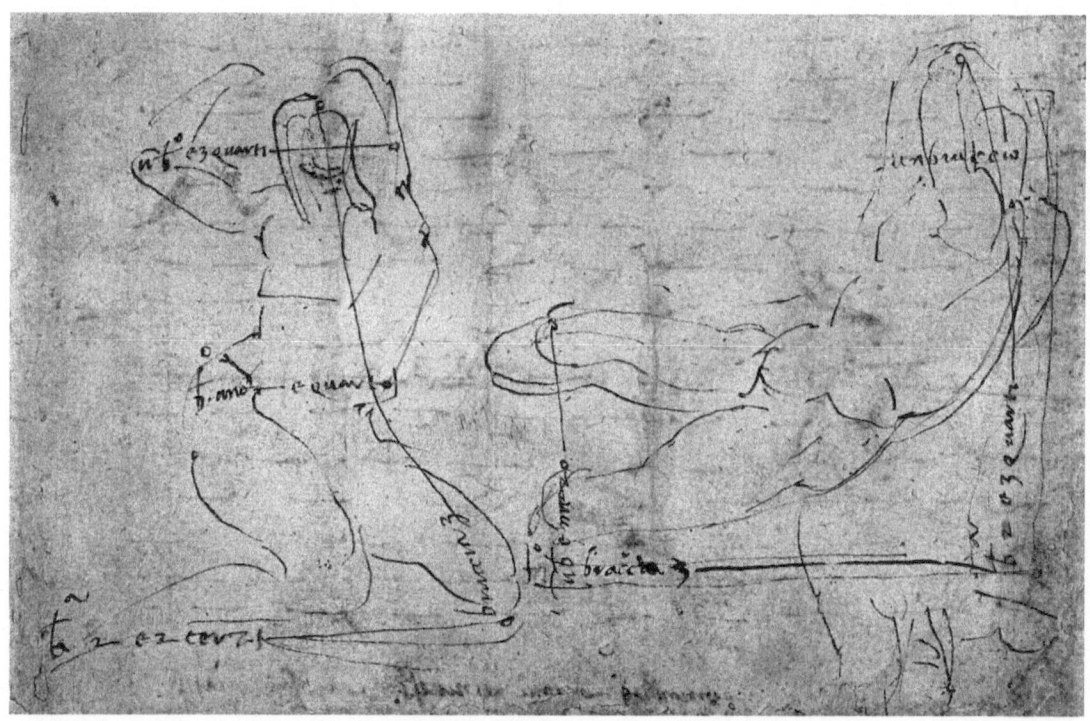

cut profile significantly deviates from the traced profile to the right, and there is no evidence along the cut line of residue from a guiding line in pencil or ink.[14] These manipulations, and the probable flipping and tracing that accompanied them, further emphasize the three-dimensionality of the paper template, an important translation in imagining the potential of paper to act as a surrogate for the projection into stone. Taking paper as a three-dimensional entity that could be cut, traced, and flipped, he discovered means for peering into the depth of stone while utilizing what was ultimately only a two-dimensional profile. Just as the distinct outline of a shadow or silhouette suggests the character of its projected body, Michelangelo's templates allowed him to penetrate into the stone body through its profile, a condition well suited for paper-like manipulations. The key analogue occurred through imagined, irreversible cuttings and through the removing of material, whether in paper or stone.

2.2   Michelangelo, sketches for the *Dei Fluviali* inside of a stone block.

This use of paper to peer deep into stone stands in stark contrast to his use of paper templates for fresco transfers, further supporting the imaginative role offered by paper as a mode of projecting stone *per forza di levare*. Although direct evidence of Michelangelo's use of cartoons in the Sistine Chapel is scant, scholars have been able to gather important clues about his transfer process.[15] During the restoration work on the Sistine Ceiling in the 1980s, scholars discovered that the artist employed a technique of tracing cartoons directly into the wet plaster using a stylus or other pointed instrument, allowing for a rapid transfer that resulted in the likely destruction of the cartoon at the same time. As well, as has been documented, this technique allowed Michelangelo to slide or reposition the cartoon on the wet plaster while in the process of tracing, a dynamic procedure that relates quite closely to his practice of shifting of tracing and shifting *modani*. The introduction of the cutting tool onto the paper edge, however, signifies a sentient alternative in how the paper is imagined. Rather than acting as a mirror-like transfer in two dimensions, paper *modani* activated the ability to conceive deep into the material and beneath the surface. By transforming the paper into a three-dimensional model, it became available as an analogue to peer into blocks of stone. Scissors, then, could be imagined as a chisel from which emerged the hiding *concetto*, deep in the material, energizing decisive reveries when bridging between paper and stone.

In defining profiles for both stone and paper, the drawing media and cutting tools were tightly bound to the artist's intentionality. Beginning with pencil, followed by ink, and continuing with the cleaving of material, the use of *modani* reflects an increasing awareness of intention, a process of self-identity more than one of artistic will. The mirror-like posture of this procedure, where the artist attacks the block or paper face-to-face, fits well into the Neo-platonic narrative of self-identity through removing the excess, a process described by Plotinus, who wrote, "Withdraw into yourself and look; do as does the maker of a statue which is to be beautiful … cut away that which is superfluous … and never cease to fashion your statue until there shall shine out upon you the godlike splendour of virtue."[16] More recently, Freud invoked the potential of *levare* as a metaphor for psychoanalysis, believing that the process of self-knowledge was akin to the psychoanalyst removing the superfluous conditions of emotional suppression.[17] Although not explicitly connected with the *via di levare*, such a procedure is also consistent with the oft-cited Renaissance maxim, *ogni dipintore dipinge se* [every painter paints himself].

In Neo-platonic terms, the multiple manifestations of the profile line, from drawing to cutting, and back to cutting again, reflects a process of self-revelation as it reciprocates between the sublunary realm and the divine intellect. The fifteenth-century Neo-platonic philosopher Marsilio Ficino, in observing that a ray of lighting will melt a piece of metal but leave its leather covering unharmed, reasons that the hardness of a material has a direct relationship with its capacity to receive divine influence. He posits that hard materials, such as metal and stone, resist heavenly rays more than softer materials, although once so inclined the harder materials are more likely to retain such influence.[18] In another example, Ficino notes that a sword striking a piece of wood wrapped in felt will likely cut the wood without damaging the wool, proving that forceful impacts may often pass through soft materials while permanently affecting those materials with the most resistance to it.[19] Under his theory of material influences, it is thus possible to retain celestial influence with soft materials, even if it is not likely; the vice-versa is also true: highly resistant materials may act as pliant receivers of divine countenance. In other words, materials have certain inclinations, but there is no fixed, stable link between material resistances and their potential for heavenly expression. Instead, Ficino offers a chiasmic relationship between material resistance and invisible influences, effectively disrupting a common understanding, even during Michelangelo's time, that harder materials are introduced later in an artist's process and thus signify greater finality.[20]

Referred to as "*Angel divino*" late in his life, Michelangelo's dexterity in wielding obdurate material with great facility upset the established order of the common artisan.[21] Stone, in this case, could be rendered as soft as flesh, an animating presence that resulted from the intersection of physical skill and the desire to release the *concetto* within.[22] The French traveler Blaise de Vigenère vividly recalled seeing the sculptor in action, instilling an image of heroic frenzy amidst the stillness of brute matter. His description is worth quoting in full:

> *I have seen Michelangelo, although more than sixty years old and no longer among the most robust, knock off more chips of a very hard marble in a quarter of an hour than three young stone carvers could have done in three or four, an almost incredible thing to one who has not seen it; and I thought the whole work would fall to pieces because he moved with such impetuosity and fury, knocking to the floor large chunks three or four fingers thick with a single blow so precisely aimed that if he had gone even minimally further than necessary, he risked loosing it all, because it could not then be repaired or re-formed, as with images of clay or plaster.*[23]

De Vigenère explicitly relates his incredulity to the heightened risk of potential loss, a key condition in erotic encounters.[24] The use of scissors, then, as an instrument of desire rather than as an indicator of finality, offered an analogous relationship with the paper that was offered to Michelangelo through the voracity of the stone chisel. By reversing the assumption that material resistances rely on conceptual finality, he activated the paper as in the stone, rendering it subtle to the imagination.

What is more, Michelangelo appears to have cut his *modani* multiple times, adjusting or tuning the profile line after several passes of the scissors. This is evident in a cornice profile destined for the upper splayed windows of the Medici Chapel, where a series of manipulations is staged as the profile emerges from the paper matrix. Following the traces of marks on both the recto and verso indicates a likely scenario where the *modano* was first traced from a now lost original, leaving black pencil residue still visible from the recto side along the template edge. The traced profile was adjusted using red chalk and ink, after which the paper was cut roughly along the inked line. It was then flipped, where on the verso a new profile line was inked. From the verso side, the template received another cut, leaving it in its current profile configuration (Figure 2.3). Thus, as was previously introduced, the paper supports the desire of an emergent figure through its ability to be treated as a three-dimensional object.

The emergence of a conception through "bit by bit" removal becomes a key catalyst for material reveries.[25] A great advantage of Michelangelo's frontal working method is that he could leave material in place to anticipate future mutations, oftentimes making significant conceptual changes through slight alterations as he went along.[26] This allowed him to continually adjust the figure narrative, engaging the reciprocity between material and its potentiality through the dissolution of any such barrier, a condition discussed earlier by Valéry. This is most evident, perhaps, in Michelangelo's *Rondanini Pietà*, left abandoned in the early 1550s but resumed

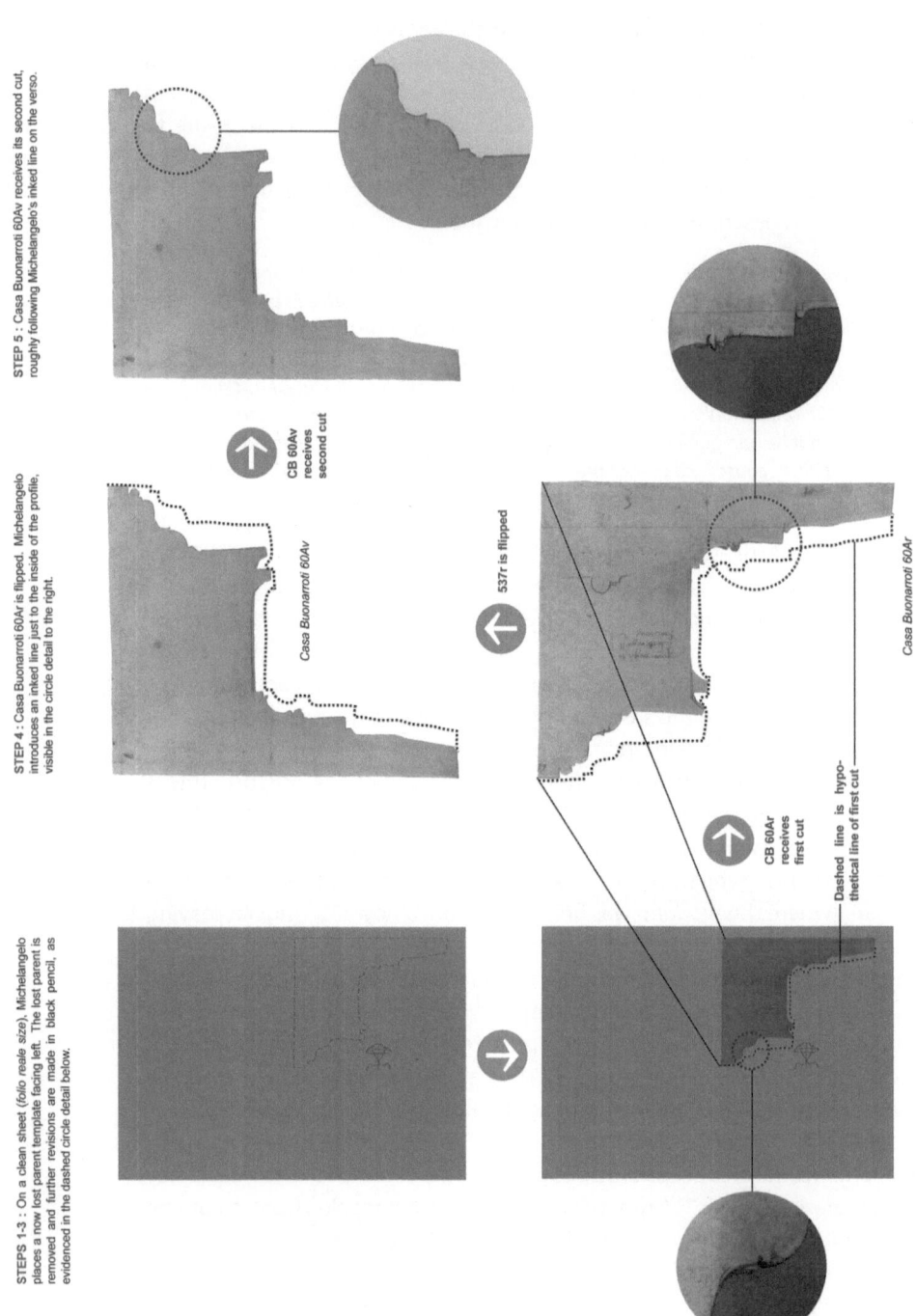

STEP 5 : Casa Buonarroti 60Av receives its second cut, roughly following Michelangelo's inked line on the verso.

STEP 4 : Casa Buonarroti 60Ar is flipped. Michelangelo introduces an inked line just to the inside of the profile, visible in the circle detail to the right.

STEPS 1-3 : On a clean sheet (folio reale size), Michelangelo places a now lost parent template facing left. The lost parent is removed and further revisions are made in black pencil, as evidenced in the dashed circle detail below.

CB 60Av receives second cut

Casa Buonarroti 60Av

537r is flipped

CB 60Ar receives first cut

Dashed line is hypo-thetical line of first cut

Casa Buonarroti 60Ar

2.3   Demonstration of manipulations for a *modano* for the Medici Chapel.

again in the final days of his life. Research has shown that he was mid-way into dramatically altering the figure group, where the Virgin's head, for example, is turning closer to Christ, and the figure of the Savior is mutating from a slouching position to standing upright. This is hardly an example of an artist proceeding uninterrupted toward a pre-conceived, mental image. Rather, the occult influence of the stone empowers an agency of material reverie, leading to an unfulfilled, rhythmic chisel that rarely relied on figural finality.

Ficino reasons, in fact, that it is the repetitive beating and heating of stone, not any conceptual pre-condition, that releases the celestial potential of material, stating that "perhaps hammering and heating alone brings out the latent power of material."[27] Repetitive, irrevocable micro-subtractions with the chisel or scissors liberated the agency of the material and locked the artist and the material within a poietic–erotic embrace, allowing for a dynamic and uninterrupted reciprocity. It is within this temporal space of chiseling fury that any conventional separation between material and artist is thoroughly disrupted. "In this way," writes Ficino in referring to clinched lovers, "they mutually exchange identities; each gives himself to the other in such a way that each receives the other in return."[28]

QUARRY DRAWINGS: EXTRACTING/SUBTRACTING

Michelangelo's *modani* are one of several drawing types that demonstrate the potential of the *forza di levare* to engage the architectural imagination. Not only was the paper a dynamic agent for his projecting the chisel into stone, he also relied on subtractive thinking by studying the stone blocks themselves through sketches and measured drawings. In this way, the reciprocity between the block and the detail was made explicit by imagining the blocks as if they were still in the quarry, waiting to emerge from their earthly matrix. From this, the rough block and the stone detail related through a dialectic that, ideally, is without both completion and origin. Following these drawings, one is not certain if the block shapes the detail or if the detail shapes the block.

In a column capital for the San Lorenzo façade, for example, Michelangelo makes a clear connection between constructing a future detail and imagining its latency within a stone block. Here, the upper drawing delineates the construction of the capital using the stylus, compass, and straightedge, a rare demonstration of Michelangelo's use of conventional architects' tools. Below, the imagined capital is postulated inside of a block of stone, labeled as such through information about its size and shape. It appears from this drawing that the column capital and its originating block were conceived simultaneously in a dynamic dialogue between formal intentions, such as dimension, orientation, and proportion, and its solid, material matrix. While it may seem at first that the block was constructed *around* Michelangelo's detail above, as a kind of additive method to imagine what would be required, further analysis of other drawings reveals that this is not so clear. Perhaps the block, previously unrelated to this detail, came first, and the drawing above denotes a possible column capital residing within it. Like the studies for the *Dei Fluviali* discussed earlier, the sequence is uncertain: did the figure give shape to

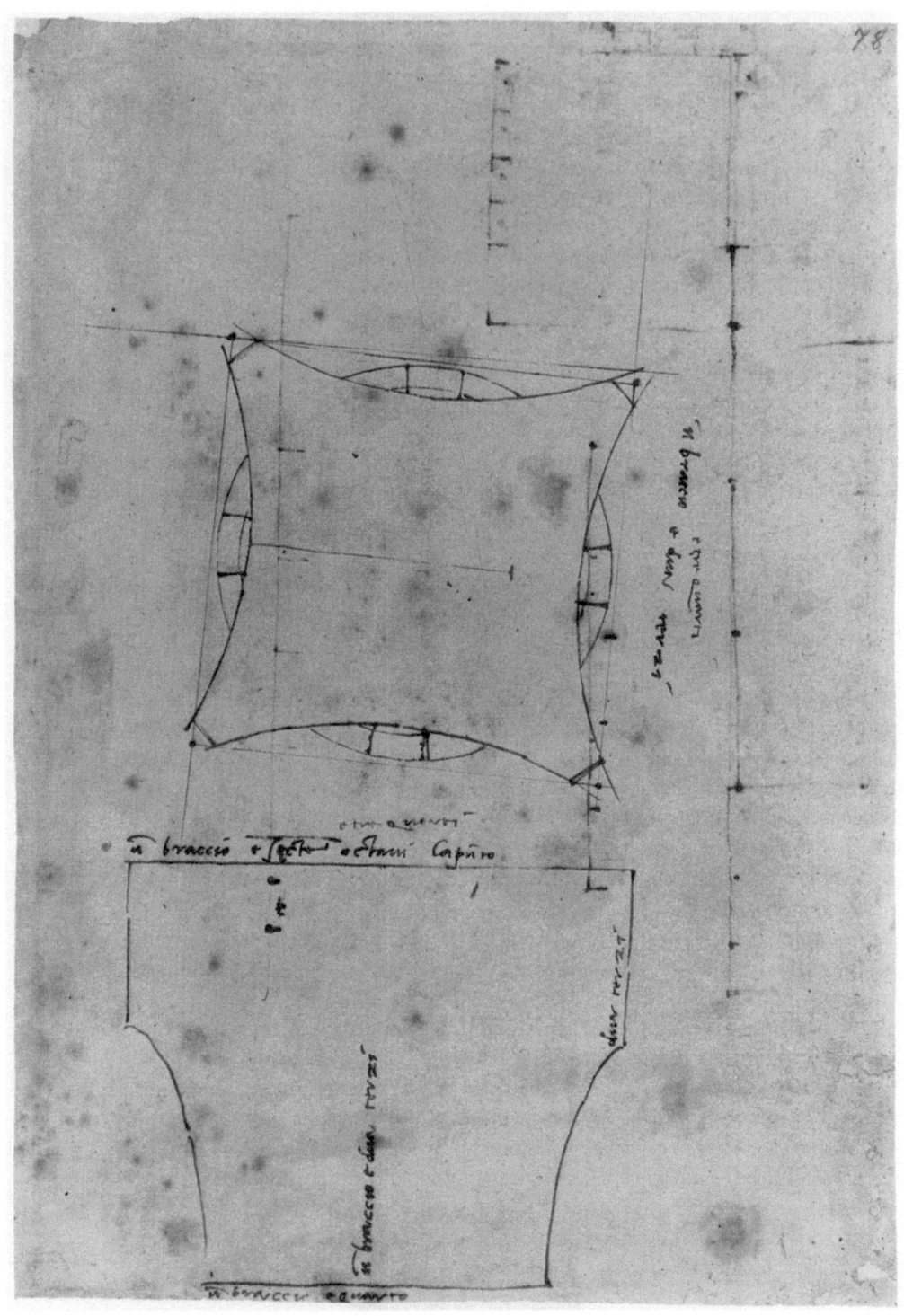

2.4   Michelangelo, column capital for San Lorenzo façade.

a hypothetical, unquarried stone block, or was a stone block present already in the workshop, waiting to be inhabited by the emerging *concetto*?

Michelangelo's reliance on massive blocks, either present or imagined, no doubt points to his nearly twenty documented trips taken to the marble quarries in the mountains around Carrara, Italy during the construction at San Lorenzo. The magnetic attraction of the white, alpine landscape provided the canvas by which the *forza di levare* inhabited the imagination, as the erotic impulse of stone certainly relied on his direct, corporal relationship with the search and extraction of stone. Just as he purposefully extracted blocks from the womb of the earth, so did he later subtract a detail or figure from a block of stone. It was here, perhaps, that he first encountered the spiritual significance of *levare*. In the quarries Michelangelo acted as a kind of midwife by removing, cultivating, and giving birth to what was embedded within the earthly matrix. The term *matrix*, in fact, was a womb in classical Latin, and it is closely related to lexical derivatives of "mater (mother)," including "material" and "matter."[29] In Italian, a midwife is called a *levatrice*, a direct derivative of *levare*. For Plato, the role of the *levatrice* was given over to Socrates, in fact, for his role in bringing forth truths latent within the youths of Athens, a point that was picked up by Ficino in his translation of *Parmenides*:[30]

> Just as Socrates, the son of a midwife, performs the office of a midwife in different places towards boys and youths and proclaims this before others, so the aged Parmenides, like a dutiful midwife, exhorts and helps the youthful Socrates to give birth to the wonderful, almost divine, opinions with which he is pregnant and which he is trying to bring forth.[31]

While we may not be sure of Ficino's exact influence on the young Michelangelo, the artist's ability to concretize these themes with remarkable consistency across his poetic and marmoreal *oeuvre* cannot be incidental.[32]

Michelangelo's quarry visits between 1516 and 1519, resulted in a series of remarkable notebooks that documented two types of stones: those that he requested from quarrymen of a certain shape, size, and number, and those that had already been removed from the earth and prepared for shipment to Florence. For each stone, Michelangelo carefully delineated its shape, dimensions, and number, and he often included information about grain direction, weight, or destination. In a drawing for one of San Lorenzo's giant, monolithic columns, for example, he sketched a general column shape, and labeled its height ("12-1/2 *braccia*"), girth ("1-1/2 *braccia*"), and number ("12").[33] Stones that had already been quarried, however, were distinguished by their lack of resemblance to recognizable elements such as lintels, column bases, or architraves. On these stones, Michelangelo also sketched his interlocking three-ring signature as what would have been carved in them as well. Altogether, the notebooks delineate some 196 stones for the San Lorenzo façade, either quarried or desired.

From these it is immediately apparent that Michelangelo developed a deep affinity with his material. What is most remarkable is that research has shown that he would often quarry more stone than he needed at the time, thus sending blocks to Florence without having a pre-determined use.[34] If he found stone of a particular quality, size, and orientation, the artist apparently extracted it based on

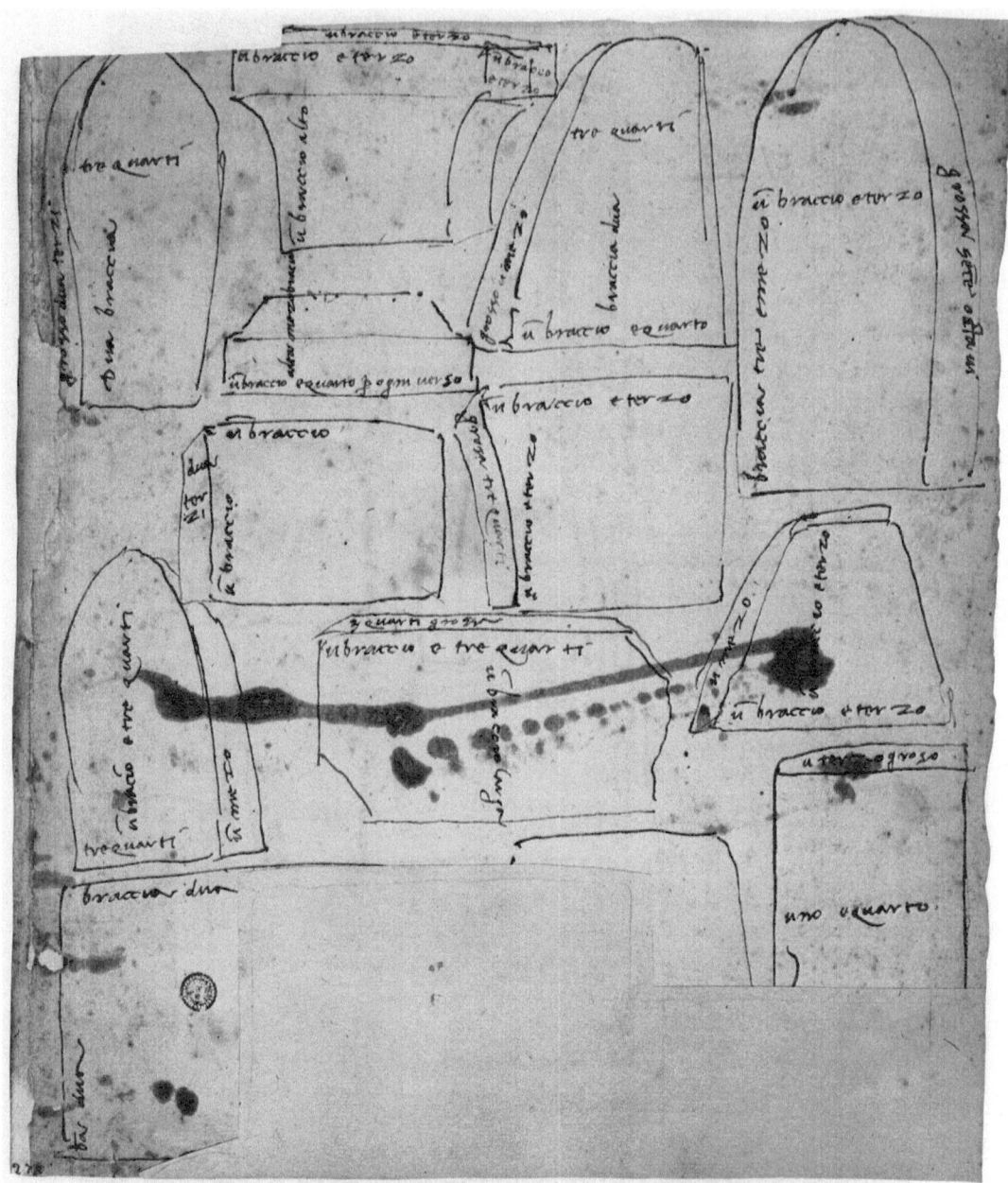

2.5   Michelangelo, quarry sketches of stone blocks.

its potential to host the desiring chisel. Peering beyond the stones' formal qualities, it seems he could imagine future *concetti* by locating stones of resonant whiteness, clarity, and grain. In this way, Michelangelo sought stones not only of certain sizes and shapes but also for patterns of hardness, veins, and directionality buried within the stone itself.[35] On a visit to the quarries of Seravezza in 1568, shortly after Michelangelo's death, the Perugian sculptor Vincenzo Danti noticed many blocks bearing the artist's carved signature that bore witness to latent *concetti* hidden within.[36] Residing in his workshop or on the building site, then, these material excesses invigorated material reveries, as they invited future details or figures to be enacted through singular presence of the material itself.

It should also be noted in these drawings the remarkable consistency in which Michelangelo imagined architectural blocks and those destined for figural works. In comparing the studies of the *Dei Fluviali* with those of the architectural blocks, the traditional sixteenth-century distinction between *lavoro di quadro* [architectural stone work] and statuary stone work has been completely blurred, something already observed in the resemblance between the *Boboli Slaves* and the frontal technique for cornice carvings.[37] Both the quarry drawings and the *Dei Fluviali* studies employ multiple views and offer equal emphasis on the potential and the actual proportions of the block, and both suggest that the formal qualities of the figure or detail could not be conceived outside of their material matrix. Michelangelo's sensory experiences in the quarries must have extracted his own stony latencies as well, a self-identified condition that originated, he once stated, by being nursed by a wife and daughter of a stonemason from Settignano.[38]

The reciprocity between material and intention found in these block drawings is just one part of a remarkably consistent pattern across Michelangelo's works, including verse, statuary, and architectural details. Coupled with the use of paper *modani* as a surrogate to stony material, he engaged the material imagination *per forza di levare* whether he worked in sculpture or architecture. In fact, throughout his projects at San Lorenzo, he continued to sign his letters, "*Michelagnolo schultore*," and did not self-identify with the discipline of architecture until he commenced work almost thirty years later on St. Peter's. The immersion in his chosen material, marble, took precedent over any conventions based on produced outcomes; rather, it was the particular reveries brought forth by the artifice that mattered. In contrast to Alberti, for example, who made clear distinctions between various kinds of sculpture, some that included techniques of addition, such as those in wax or clay, Michelangelo offered a definition of sculpture based on removal only.[39] Among a low point in the artist's life, in fact, was when he was pressed into service by Pope Julius II to lead a monumental bronze casting effort of his likeness in Bologna. This work, which was destroyed by mobs shortly after completion, remains a testament to Michelangelo's general dislike of casting and molding, both as a practical technique as well as its potential as a metaphor for transcendence.[40]

Whether envisioning the work with his own hands or through the hands of others, his catalyst for the imagination was the fundamental sculptural encounter. Writing in 1525 at the height of his work at San Lorenzo, Michelangelo wrote, "one cannot shape one thing with one's hands, and another with one's brain, especially in marble."[41] This statement, a confirmation of the artist's unity of mind with the hand, is curious considering that Michelangelo was not documented to have carved any of the architectural stonework himself at San Lorenzo. Instead of confronting the stone directly, he developed a complex method of using paper as surrogate for his imagined matrix through tracing and cutting—methods that were governed by the conceptual imperative of *levare*. The barrier between the artist and his assistants seems additionally blurred, as nearly all of the most active stone carvers on the San Lorenzo building site were either life-long assistants or fellow carver acquaintances from around the hills of Settignano, his birthplace. At the same time, as has been shown in analyzing his *modani*, Michelangelo could think in stone as if we were carving it himself, opening up the possibility to think figurally about details rather than in rigid, mathematical modules. While much has been written about the artist's formal choices and their possible origins in pre-determined narratives, the material imperative and the *forza di levare* and accompanying reveries cannot be overlooked.

BIBLIOGRAPHY

Alberti, Leon Battista, Cosimo Bartoli, and Girolamo Tiraboschi, *Della pittura e della statua di Leonbatista Alberti* (Milan: Società tip. de'Classici italiani, 1804).

Bachelard, Gaston, *Earth and Reveries of Will: An Essay on the Imagination of Matter* (Dallas, TX: Dallas Institute of Humanities and Culture, 2002).

Bambach, Carmen, *Drawing and Painting in the Italian Renaissance Workshop: Theory and Practice, 1300–1600* (Cambridge, MA: Cambridge University Press, 1999).

Barocchi, Paola, *Scritti d'arte del Cinquecento: a cura di Paola Barocchi* (Milan: Ricciardi, 1971).

Barocchi, Paola, and Renzo Ristori, *Il carteggio di Michelangelo* (Florence: Sansoni, 1979).

Carabell, Paula, "Image and Identity in the Unfinished Works of Michelangelo," *RES: Anthropology and Aesthetics*, no. 32 (Autumn 1997), pp. 83–105.

Cellini, Benvenuto, and Carlo Milanesi, *I Trattati dell'oreficeria e della scultura* (Florence: Felice Le Monnier, 1857).

Clements, Robert J., *Michelangelo's Theory of Art* (New York, NY: New York University Press, 1961).

Cooper, Tracy E., "I Modani: Template Drawings," in Henry A. Millon and Vittorio M. Lampugnani (eds), *The Renaissance from Brunelleschi to Michelangelo* (Milan: Bompiani, 1994).

Elam, Caroline, "Funzione, tipo e ricezione dei disegni di architettura di Michelangelo," in C. Elam (ed.), *Michelangelo e il disegno di architettura* (Venice: Marsilio, 2006).

Ficino, Marsilio, and S.R. Jayne, *Marsilio Ficino's Commentary on Plato's Symposium* (Columbia, MO: University of Missouri, 1944).

Ficino, Marsilio, and Maude Vanhaelen, *Commentaries on Plato: Parmenides* (Cambridge, MA: Harvard University Press, 2012).

Ficino, Marsilio, Carol V. Kaske, and John R. Clark, *Three Books on Life* (Binghamton, NY: Medieval & Renaissance Texts & Studies in conjunction with the Renaissance Society of America, 1989).

Freud, Sigmund, *Selected Papers on Hysteria and Other Psychoneuroses*, trans. A.A. Brill (New York, NY: The Journal of Nervous and Mental Disease Publishing Company, 1912).

Hub, Berthold, "Material Gazes and Flying Images in Marsilio Ficino and Michelangelo," in Christine Göttler and Wolfgang Neuber (eds), *Spirits Unseen: The Representation of Subtle Bodies in Early Modern European Culture* (Boston, MA: Brill, 2008).

Lavin, Irving, "David's Sling and Michelangelo's Bow: A Sign of Freedom," in *Past–Present: Essays on Historicism in Art from Donatello to Picasso* (Berkeley, CA: University of California Press, 1993).

Milanesi, Gaetano, *Le Lettere di Michelangelo Buonarroti. Pubblicate coi ricordi ed i contratti artistici per cura di Gaetano Milanesi* (Florence: Le Monnier, 1875).

Paul, Joannides, "Catalogue Entry to Casa Buonarroti 59A and 61A," in Serjez O. Androsev and Umberto Baldini (eds), *L'Adolescente dell'Ermitage e la Sagrestia Nuova di Michelangelo* (Florence: Artout, Maschietto & Musolino, 2000).

Saslow, James, *The Poetry of Michelangelo: An Annotated Translation* (New Haven, CT: Yale University Press, 1991).

Summers, David, *Michelangelo and the Language of Art* (Princeton, NJ: Princeton University Press, 1981).

Taylor, Thomas, *Concerning the Beautiful, or a Paraphrased Translation from the Greek of Plotinus. Ennead I. Book VI. by T. Taylor* (1787).

Tolnay, Charles De, *Corpus dei disegni di Michelangelo* (Novara: Istituto geografico De Agostini, 1975–1980).

Valéry, Paul, *Degas, Manet, Morisot* (New York, NY: Pantheon, 1960).

Vasari, Giorgio, and Gaetano Milanesi, *Le vite de' più eccellenti pittori, scultori ed architettori* (Florence: G.C. Sansoni, 1906), vol. I.

Wallace, William E., *Michelangelo at San Lorenzo: The Genius as Entrepreneur* (Cambridge: Cambridge University Press, 1994).

Wallace, William E., "Michelangelo's Wet Nurse," *Arion*, Third Series, vol. 17, no. 2 (2009): pp. 51–5.

NOTES

1   "… fu necessario trovar qualcosa poi di marmo, perchè e' potessi ogni giorno passar tempo scarpellando," Giorgio Vasari and Gaetano Milanesi, *Le vite de' più eccellenti pittori, scultori ed architettori* (Florence, 1906), vol. VII, pp. 244–5.

2   For an overview of the *paragone* debate of the sixteenth century and Michelangelo's contributions, see David Summers, *Michelangelo and the Language of Art* (Princeton, 1981), pp. 273–90.

3   "Io intendo scultura quella che si fa per forza di levare; quella che si fa per via di porre è simile a la pictura," Paola Barocchi and Renzo Ristori, *Il carteggio di Michelangelo* (Florence, 1979), vol. IV, pp. 265–6.

4   Leonardo emphatically stated the painting required greater mental difficulty, even if sculpture was more difficult for the body, and thus painting emerges superior over sculpture, Paola Barocchi, *Scritti d'arte del Cinquecento: a cura di Paola Barocchi* (Milan, 1971), pp. 484–5.

5   "Sì come per levar, donna, si pone / in pietra alpestra e dura / una viva figura, / che là più cresce u' più / la pietra scema; / tal alcun'opre buone, / per l'alma che pur trema, / cela il superchio della propria carne / co' l'inculta sua cruda e dura scorza / Tu pur dalle mie streme / parti puo' sol levarne, / ch'in me non è di me voler né forza," translation and transcription from James Saslow, *The Poetry of Michelangelo: An Annotated Translation* (New Haven, 1991), no. 152, p. 305.

6   Paul Valéry, *Degas, Manet, Morisot* (New York, 1960), p. 144.

7   A good starting point is Paula Carabell, "Image and Identity in the Unfinished Works of Michelangelo," *RES: Anthropology and Aesthetics*, no. 32 (Autumn 1997), pp. 83–105.

8   Robert J. Clements, *Michelangelo's Theory of Art* (New York, 1961), p. 24.

9   Extant modani from the cinquecento are rare, and none are known to exist from the fifteenth century and earlier. For an introduction, see Tracy E. Cooper, "I Modani: Template Drawings," in Henry A. Millon and Vittorio M. Lampugnani (eds), *The Renaissance from Brunelleschi to Michelangelo* (Milan, 1994), pp. 494–9.

10   "Et il miglior modo che si sia mai visto è quello che ha usato il gran Michelagnolo, il qual modo si è di poi che uno ha disegnato la veduta principale, si debbe per quella banda cominciare a scoprire con la virtù de' ferri, come se uno volessi fare una figura di messo rilievo, e così a poco a poco si viene scoprendo," Benvenuto Cellini and Carlo Milanesi, *I Trattati dell'oreficeria e della scultura* (Florence, 1857), ch. VI, p. 198. Translation by author.

11   "… prima scoprendo le parti più rilevate, e di mano in mano le più basse: il quale modo si vede osservato da Michelagnolo ne'sopradetti prigioni …," Vasari, vol. VII, p. 273.

12   "Non ha l'ottimo artista alcun concetto / c'un marmo solo in se' non circonscriva / col suo superchio, e / solo a quallo arriva / la man che ubbidisce all'intelletto," translation and transcription by Saslow, no. 151, p. 302.

13   Joannides Paul observed that the cutting with scissors constituted a "prosecuzione del processo creativo," Joannides Paul, "Catalogue Entry to Casa Buonarroti 59A and 61A," in Serjej O. Androsev and Umberto Baldini (eds), *L'Adolescente dell'Ermitage e la Sagrestia Nuova di Michelangelo* (Florence, 2000), pp. 132–4.

14   Charles De Tolnay, *Corpus dei disegni di Michelangelo* (Novara, 1975–1980), vol. IV, no. 536.

15   For an introduction to Michelangelo's use of cartoons, see Carmen Bambach, *Drawing and Painting in the Italian Renaissance Workshop: Theory and Practice, 1300–1600* (Cambridge, 1999), p. 133 and p. 420, note 28.

16   Plotinus, *Ennead* 1.6.IX, translated by Thomas Taylor, *Concerning the Beautiful; or a paraphrased translation from the Greek of Plotinus. Ennead I. Book VI* (1787). The origin of Neo-platonic thought in this regard is often sourced to Aristotle, *Metaphysics*, IX, 6, where he discusses the difference between actuality and potentiality. See also Marsilio Ficino and S.R. Jayne, *Marsilio Ficino's Commentary on Plato's Symposium* (Columbia, 1944), Sixth Speech, ch. XVII.

17   Sigmund Freud, *Selected Papers on Hysteria and Other Psychoneuroses*, trans. A.A. Brill (New York, 1912), vol. VIII, p. 9.

18   Marsilio Ficino, Carol V. Kaske, and John R. Clark, *Three Books on Life* (Binghamton, 1989), p. 323.

19   Ibid.

20   Vasari describes the careful progression of conventional carving methods in Giorgio Vasari and Gaetano Milanesi, *Le vite de' più eccellenti pittori, scultori ed architettori* (Florence, 1906), vol. I, p. 155.

21   This reference is found in Ludovico Ariosto, *Orlando furioso*, XXXIII.2.4, a popular poem published in 1532.

22   Ovid describes melting stone by the hand of the artist who falls in love with his statue in Carabell, p. 90.

23   Quoted from Irving Lavin, "David's Sling and Michelangelo's Bow: A Sign of Freedom," in *Past–Present: Essays on Historicism in Art from Donatello to Picasso* (Berkeley, 1993), p. 43.

24   See also Ficino, *Symposium*, Second Speech, ch. VI.

25   Vasari describes this as "di mano in mano" in Vasari, vol. I, p. 155. Antonio Manetti describes how Brunelleschi dictated his detail to the craftsmen "cosa per cosa" in his biography of the architect.

26   Vasari describes this condition of Michelangelo's method in Vasari, vol. VII, pp. 157–8.

27   Ficino, Kaske, and Clark, *Three Books on Life*, p. 343.

28   Ficino, *Symposium*, Second Speech, ch. VIII, p. 144.

29   "matrix, n.". *Oxford English Dictionary*, Online Dictionary, Oxford University Press, 2014. Available at: www.oed.com [accessed August 26, 2014].

30   The Greek term for Socrates' practice was "*maieutic*," or "obstetrics."

31   Marsilio Ficino and Maude Vanhaelen, *Commentaries on Plato: Parmenides* (Cambridge, 2012), ch. 16.

32   On an introduction to Ficino and Michelangelo, see Berthold Hub, "Material Gazes and Flying Images in Marsilio Ficino and Michelangelo," in Christine Göttler and Wolfgang Neuber (eds), *Spirits Unseen: The Representation of Subtle Bodies in Early Modern European Culture* (Boston, 2008), pp. 93–120.

33   De Tolnay, vol. IV, no. 442.

34   William E. Wallace, *Michelangelo at San Lorenzo: The Genius as Entrepreneur* (Cambridge, England, 1994), pp. 42–4.

35   Michelangelo comments on marble qualities in several places, for example: "… senza alcuni peli," in Gaetano Milanesi, *Le Lettere di Michelangelo Buonarroti. Pubblicate coi ricordi ed i contratti artistici per cura di Gaetano Milanesi* (Florence, 1875); and "… vale marmi mostrono di volere mostrare la luna ne' pozo," Barocchi and Ristori, vol. II, p. 6. Following this, the philosopher Gaston Bachelard reflects that sculptors read materials such as stone through variations in hardness, "there are sculptors who use hardness in the same way that painters use pigment," Gaston Bachelard, *Earth and Reveries of Will: An Essay on the Imagination of Matter* (Dallas, 2002), p. 40.

36   Barocchi and Ristori, vol. III, p. 258.

37   On *lavoro di quadro*, see Caroline Elam, "Funzione, tipo e ricezione dei disegni di architettura di Michelangelo," in *Michelangelo e il disegno di architettura* (Venice, 2006), p. 45, and Vasari, vol. I, p. 29.

38   See William E. Wallace, "Michelangelo's Wet Nurse," *Arion*, Third Series 17, no. 2 (2009), pp. 51–5.

39   Leon Battista Alberti, Cosimo Bartoli, and Girolamo Tiraboschi, *Della pittura e della statua di Leonbatista Alberti* (Milan, 1804), pp. 107–8.

40   One exception is a love sonnet found in Saslow, p. 306, although here the emphasis is on the heat of the forge, and less on metal as such.

41   "… che e' non si puo lavorare con le mani una cosa, e col ciervello una altra, e massimo di marmo," Barocchi and Ristori, vol. III, pp. 173–4.

3

Phenomena and Idea[1]

Steven Holl

The original text was written in 1992, twenty-two years ago. Rereading it today I would remove the word "phenomenology."

Steven Holl, April 23, 2014

Experience of phenomena—sensations in space and time as distinguished from the perception of objects—provides a "pre-theoretical" ground for architecture. Such perception is pre-logical, i.e. it requires a suspension of a priori thought. Phenomenology, in dealing with questions of perception, encourages us to experience architecture by walking through it, touching it, listening to it. "Seeing things" requires slipping into a world below the everyday neurosis of the functioning world. An underground city for which we have keys without locks, it is full of mysteries.

3.1   Light studies for the Museum of the City, Cassino, Italy, 1996.

Phenomenology as a way of thinking and seeing becomes an agent for architectural conception. While phenomenology restores us to the importance of lived experience in authentic philosophy, it relies on perception of pre-existing conditions. It has no way of forming a priori beginnings. Making a non-empirical architecture requires a conception or a formative idea. In each project we begin with information and disorder, confusion of purpose, program ambiguity, an infinity of materials and forms. All of these elements, like obfuscating smoke, swirl in a nervous atmosphere. Architecture is a result of acting on this indeterminacy.

To open architecture to questions of perception, we must suspend disbelief, disengage the rational half of the mind, and simply play and explore. Reason and skepticism must yield to a horizon of discovery. Doctrines cannot be trusted in this laboratory. Intuition is our muse. The creative spirit must be followed with happy abandon. A time of research precedes synthesis.

In music one says that something is "meant" by a particular movement. Do architectural thoughts have equivalent "meanings"? Is there a way of thinking in the material of construction? A way of thinking in material which may yield a coupling of thinking–making specific to architecture? Making architecture involves a thought that forms itself through the material in which it is made. The thinking–making couple of architecture occurs in silence. Afterward, these "thoughts" are communicated in the silence of phenomenal experiences. We hear the "music" of architecture as we move through spaces while arcs of sunlight beam white light and shadow.

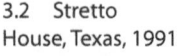

3.2   Stretto House, Texas, 1991.

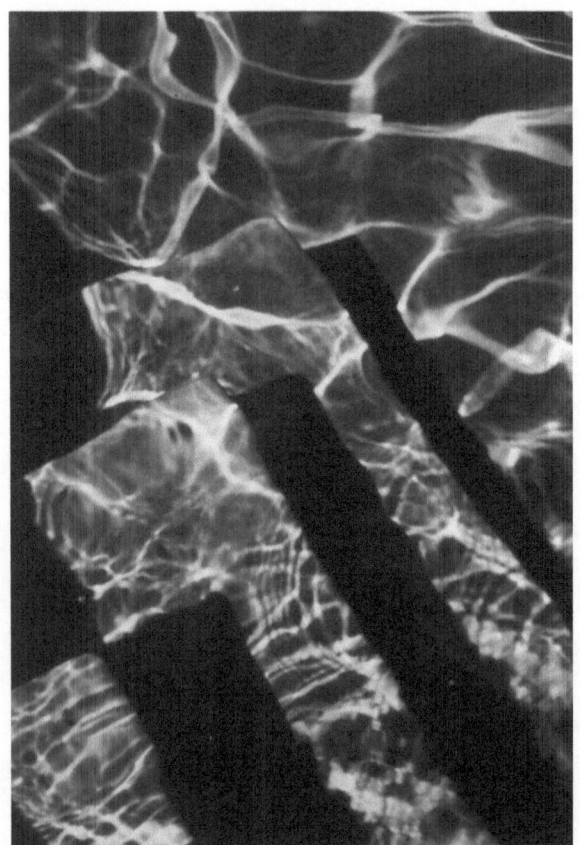

In a "zero ground" without site, program, or time, certain types of perception emerge as "phenomenal zones." Experimental territories, these zones of intensely charged silence lie beyond words. In opposition to those who insist on speech, on language, on signs and referents, we strive to escape language–time bondage. To evolve theoretically in active silence encourages experimentation. Silent phenomenal probes haunt the polluted sea of language like submarines gliding along the sandy bottom, below the oil-slick of rhetoric.

Certain physical interactions offer zones of investigation:

Color projection is experienced when light, reflected off a brightly colored surface, then bounced onto a neutral white surface, becomes a glowing phenomena that provokes a spatial sense. Reflected color is seen indirectly; it remains, with a ghostlike blush, the absent referent to an experience.

In experiments with these phenomena we have discovered an emotional dimension that suggests a "psychological space."

A sponge can absorb several times its weight in liquid without changing its appearance. Cast glass seems to trap light within its material. Its translucency or transparency maintains a glow of reflected light, refracted light, or the light dispersed on adjacent surfaces. This intermeshing of material properties and optic phenomena opens a field for exploration. Phenomenal zones likewise open to sound, smell, taste, and temperature as well as to material transformation.

Overlapping perspectives, due to movement of the position of the body through space, create multiple vanishing points, opening a condition of spatial parallax. Perspectival space considered through the parallax of spatial movement differs radically from the static perspectival point of Renaissance space and the rational positivist space of modern axonometric projection. A dynamic succession of perspectives generates the fluid space experienced from the point of view of a body moving along an axis of gliding change. This axis is not confined to the x–y plane but includes the x–y–z dimensions manifesting themselves in other dimensions, gravitational forces, electromagnetic fields, time, etc. Perspectives of phenomenal flux, overlapping perspective space is the "pure space" of experiential ground.

Architecture is born when actual phenomena and the idea that drives it intersect. Whether a rationally explicit statement or a subjective demonstration,

3.3 Nelson–Atkins Museum of Art, Kansas City, MO, 2007.

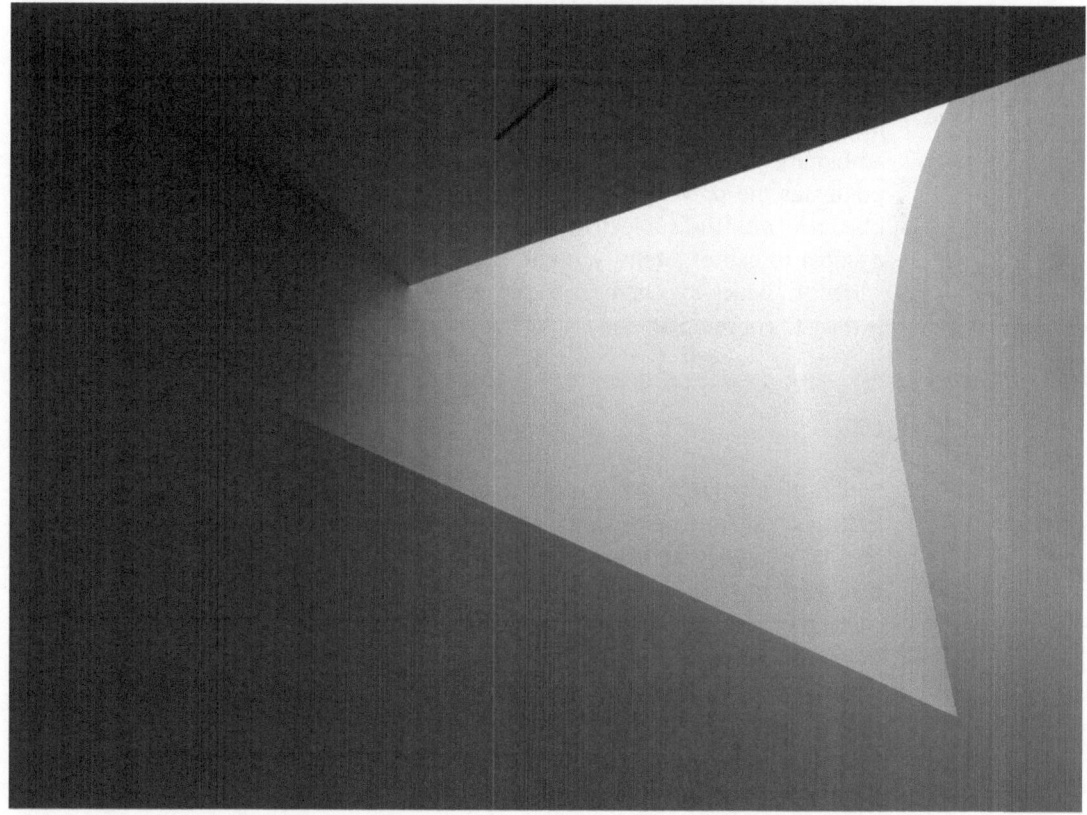

a concept establishes an order, a field of inquiry, a limiting principle. The concept acts as a hidden thread connecting disparate parts with exact intention. Meanings show through at this intersection of concept and experience.

A structuring thought requires continuous adjustment in the design process to set manifold relations among parts within the larger whole. As dimensions of perception and experience unfold in the design process, constant adjustments aim at a balance of idea and phenomena.

"KAJITSU"

Japanese Zen poets developed a vocabulary to discuss *Kajitsu* of a poem's aspect and form. *Ka* is the beautiful surface of a poem while *jitsu* is the substantial core. An organic fusion of spirit and intellect opens a path toward inspiration, awareness, and *yugen*, the Buddhist term for "depth of meaning."

Uncovering the elusive essence of architecture, its depth of meaning or substantial core, requires passion and enthusiasm. The search for meaning demands a resistance to empty formalism, textual obfuscation, and commercialism. Focusing on ideas early in the design process sets the substantial core ahead of the surface.

If there is a life in ideas, a passion for architecture is renewed in the clarification of these ideas. For what is an architectural concept if not the material and spatial expression of spiritual intentions?

Intertwining of intellect and feeling is inherent in thought intuitively developed, thought that seeks clarity rather than possesses truth, thought that searches and is open to the changing field of culture and nature that it expresses. Although intuition cannot be explicitly expressed, we cannot condemn intuitive work to ambiguity. Architecture perhaps more than any other form of communication, possesses the power of uniting intellectual and intuitive expression. Fusing the objective with the subjective, architecture can stitch our daily lives together by a single thread of intensity. It can possess both the core depth and the radiant surface by which to concretize the spirit. We must look beyond the *ka* of a beautiful surface to contemplate the *jitsu* of the core substance.

SOUL

Soul is essential to architecture. A building stands in mute solitude, yet receptive individuals silently perceive the soul instilled in the work. Soul lies in attention to detail distilled in space and concretized in the love of construction. This love can take the form of shimmering icicle prisms or perspectives of steel.

In the thirteenth century, Saint Thomas Aquinas developed teachings linking theology and philosophy which held that all knowledge begins with sense perception. The direct connection of soul and perception was taught in "clear sighted penetration of the soul into objects of perception ...".

Nourishment of soul begins by allowing greater expression of the language of the imagination, by suspending belief in favor of experiment, and by seeing things.

3.4   T-Space, Dutchess County, New York, 2010.

Cultivating of a metaphorical sense of reality … a mythopoetic understanding of indefinable experiences and mysteries enriches the soul. Just as the unconscious and the intuitive can be intentionally brought to bear on thoughts and decisions, the intense exploration of a particular locus together with material, can endow form with greater psychological significance. Like an electrical charge, soul passes from the artist into objects, and through eyes from the object to the viewer. Art is like a battery holding that charge for the future.[2]

Reflection on perception in the design process considers all scales, including the micro scale of material properties. Even the most common seemingly inert material must be allowed to "speak" its essence. Kandinsky addresses this approach: "Everything that is dead quivers. Not only the things of poetry, stars, moon, wood, flowers, but even a white trouser button glittering out of a puddle in the street. Everything has a secret soul, which is silent more than it speaks."

Triumphant expressions of life often emerge despite the cycles of death by which they are surrounded. The question of soul is a question of will. The spirit of a community or society as well as that of an individual is often a pathologized territory. New investigations and new projects must be undertaken. Today the urgency of the soul is provoked by unprecedented human coldness.

An inexplicable modern soul unfolds from tragedy and absurdity. Hope rises on the ground of desperate conditions indirectly proportional to the emotional intensity of the situation: in the writings of Franz Kafka, and André Breton, the tragic and seemingly absurd are taken to extremes, yielding a strange existential hope. Humiliating circumstances and absurd predicaments are part of everyday life in the modern metropolis, yet these conditions fuel the modern soul.

To embrace the unique anxieties of our time, one must avoid false optimism and the phantoms of nostalgia. Our challenge is to make spaces of a serenity and exhilaration that allow the modern soul to emerge. Our everyday lives include the upside-down view of the earth, in a live television broadcast in which figures walk without gravity, or stroll along a sidewalk past barrels of live crabs fighting each other. The modern soul, its unprecedented spirit, must have an architecture.

MESHING SENSATION AND THOUGHT

> If we walk along a shore towards a ship which has run aground, and the funnel or masts merge into the forest bordering on the sand dune, there will be a moment when these details suddenly become part of the ship; and indissolubly fused with it. As I approached, I did not perceive resemblances or proximities which finally came together to form a continuous picture of the upper part of the ship. I merely felt that the look of the object was on the point of altering, that something was imminent in this tension, as a storm is imminent in storm clouds.
>
> Maurice Merleau-Ponty

Perception of architecture entails manifold relations of three fields; the foreground, middle ground, and distant view are united in one experience as we observe and reflect while occupying a space. Mergings of these fields of space bracket very different perceptions. In the intertwining of the large space with its forms and proportions and the smaller scale of materials and details lies architecture's power to exhilarate. Such phenomenal territory cannot be indicated in plan/section methods. Photography can only present one field clearly, excluding changes in space and time.

The weak link from perception back to inception must be scrutinized and strengthened. The traditional drawing of a plan is a blind notation, nonspatial and nontemporal. Perspectives of overlapping fields of space break this short circuit in the design process. Perspective precedes plan and section to give a priority to bodily experience and binds creator and perceiver. The spatial poetry of movement through overlapping fields is animated parallax.

To work simultaneously in foreground, middle ground, and distant view, an architect must constantly think of the next smaller and the next larger scales. The master plan of a campus space, for example, must consider the space between and within buildings as well as details of materials, glossy or dull or luminescent. Models constructed in plaster, wire, acid-transformed brass, and other construction materials balanced against a range of perspective views set an intermeshing design process in motion.

The phenomenal merge of object and field is accomplished via attention to individual site and situation. The hackneyed terms *contextualism* or *context* have encouraged an operation whereby a new building, chameleon like, takes characteristics from each of its neighbors without maintaining internal integrity.

Rather, actual experience envisioned in light, perspective, and material must be cross-referenced in an analytic process open to a new architecture that may not yet be understood. Architecture inserted into an existing situation may not strive to replicate or to achieve autonomy via contrast. Meshing of site and situation with an integrally conceived new architecture yields a third condition; a new interrelation—a new "place" is formed.

TIME'S MULTIPLICITY

As the imperceptible downward flow of glass in the lower portion of window panes measures the passage of time, architecture also serves as an index of time. Second, minute, hour, month, year, decade, epoch, millennium all are focused by the lens of architecture. Architecture is among the least ephemeral, most permanent expressions of culture.

Nostalgia, an irrational yearning for the return to another time, dominates American architecture today. Preservation of the past continues in the mind, in books, in photographs and films, and in the conservation of past construction, however simulating the past is a travesty of the present.[3] This return to a romanticized time avoids the existential burden of time—its angst and its joy.

A certain resistance, a "negative capability," is necessary to exist and act in the present. It is important to think and to act on our thinking in the present. We are not merely of our time. We *are* our time. In our time the nature of speed itself has transformed the definition of space. The acceleration of fluctuating trends renders it impossible to meet ever-changing appetites. To last, to endure, is a primary challenge of architecture conceived today.[4]

Strategies transcending the novel and image driven in architecture counter the ongoing historical time of Western culture with a cyclical time of particular place and individual circumstance. For each distant situation there is a time, yielding a "multiplicity of times." For example, for Islamic theologians time is not a continuous flow but a galaxy of instants. Space is nonexistent except in points. Alternately, Bergson's idea of "duration" includes a "multiplicity of secession, fusion and organization."[5] These two ideas of time—as space or as continuous multiplicity and flow—correspond to the strange cultural conditions of the world today.

While a global movement electronically connects all places and cultures in a continuous time–place fusion, the opposite tendency coexists in the uprising of local cultures and expression of place. In these two forces—one a kind of expansion, the other a kind of contraction—time–space is being formed. A new architecture must be formed that is simultaneously aligned with transcultural continuity and with a poetic expression of individual situation and community.

Expanding toward an ultra-modern world of flow while condensing sunlight or the texture of stone, on a single plot of land, this architecture aspires to Blake's admonition "to see the universe in a grain of sand." Poetic illumination of unique qualities of places, individual culture and individual spirit reciprocally connects to the transcultural, trans-historical present.

Architecture is a transforming link. An art of duration, crossing the abyss between ideas and orders of perception, between flow and place, it is a binding force, It bridges the yawning gap between the intellect and senses of sight, sound, and touch, between the highest aspirations of thought and the body's visceral and emotional desires. A multiplicity of times are fastened, a multitude of phenomena are fused, and a manifold intention is realized.

IDEA

It is precisely the realm of ideas—not forms or styles—that presents the most promising legacy of twentieth-century architecture. The twenty-first century propels architecture into a world where meanings cannot be completely supplied by historical languages. Modern life brings with it the problem of the meaning of the larger whole. The increased size and programmatic complexity of buildings amplify the innate tendency of architecture toward abstraction. The tall office buildings, the urban apartment house, and the hybrid of a commercial complex call for more open ideas, more imaginative organization of a work of architecture. Organization of overall form depends on a central concept to which other elements remain subordinate.

In the experimental work of tentative investigations we remain explorers. This new freedom produces an anxiety that must be embraced with enthusiasm. The practice of a refined methodology, a technical skill, has now seeped through the osmotic membrane of a narrow profession into the open sea and must be

nourished with a passion for discovery. New architecture can only be born if we leave habitual ways of working and reject unthinking methods.

Easily grasped images are the signature of today's culture of consumer architecture. Subtle experiences of perception as well as intellectual intensity are overshadowed by familiarity. A resistance to commercialism and repetition is not only necessary, it is essential to a culture of architecture.

The experience of space, light, and material as well as the socially condensing forces of architecture are the fruit of a developed idea. When the intellectual realm, the realm of ideas, is in balance with the experiential realm, the realm of phenomena, form is animated with meaning. In this balance, architecture has both intellectual and physical intensity, with the potential to touch mind, eye, and soul.

3.5   Nelson–Atkins Museum of Art, Kansas City, MO, 2007.

NOTES

1   Originally published in *GA Architect 11: Steven Holl*, ed. Y. Futagawa (1992), pp. 12, 16–17.

2   This sentence is not in the original published text. Added by Steven Holl.

3   The original published text reads, "Preservation of the past continues in the mind in books, in photographs and films, and in the conservation of past construction but simulating the past is a travesty of the present." Changes by Steven Holl.

4   The original published text reads, "To last, to endure, is a primary challenge to architecture conceived today." Changes by Steven Holl.

5   The original published text reads, "Alternately, from Bergson's idea of 'duration' includes a 'multiplicity of secession, fusion and organization.'" Changes by Steven Holl.

4

The Resistance of Factures in Drawing-out Architectural Constructions

Matthew Mindrup

> *The [drawn] line no longer imitates the visible; it "renders visible"; it is the blueprint of a genesis of things.*[1]
>
> Maurice Merleau-Ponty

By inscribing a line on paper, the architect vivifies mute material and transforms a flat surface into a virtual space. English dictionaries use words such as act, art, and image to define "drawing." However, with each additional line, an architect also experiences a second sense of "drawing" as a "pulling out" or revealing of a desired internal idea. At the beginning of a new project, the image of a proposed place is usually incomplete and design involves the different methods architects employ to make ideas visible. The art historian and critic James Elkins has observed a similar exchange between the external and internal image in his own work, suggesting how:

> *[e]very line I draw reforms the figure on the paper, and at the same time it redraws the image in my mind. And what is more, the drawn line redraws the model, because it changes my capacity to perceive.*[2]

Elkin's remark reveals a key characteristic about the act of drawing. Each additional mark on an empty surface of paper records an act of intention but also invites the architect to enter into dialectic with the emerging graphic construction and their mental image. In this way, drawing by hand plays an important role in the architect's *oeuvre* as a tool both to record a preconceived idea and to aid in the construction of one.

Architects draw buildings using graphic processes to explore physical acts of construction and experience. An architect mixes personal and conventional methods for describing architectural elements, the primary aim of drawing being to foresee and understand a proposed project. In a statement about his design process, the Italian architect Carlo Scarpa affirms this application of drawing in his own work, suggesting that is it because "I want to see things, that's all I really trust. I

4.1 Peter Zumthor, Preliminary Pastel "Block Drawing" for Thermal Bath House, Vals, Switzerland, 1990.

want to see, and that's why I draw. I can see an image only if I draw it."[3] Because the act of drawing is in many ways dependent upon memory, an architect's previous thoughts, images, and experiences strongly influence what will be put on paper. This implies that architects who desire to develop new designs must employ a method of drawing to take them beyond preconceived ideas and standardized approaches.

As indexical signs for the designation of building elements, architectural drawings can be more than an aggregate of conventional traces denoting pre-determined design decisions. Instead, intentional marks, accidents, and incomplete gestures can leave "open patches" in the delineation of a project that engender the architect's imagination and invite creative speculation. The Swiss architect Peter Zumthor ascribes a considerable amount of importance to the imperfection of drawing as a tool to help the architect discover new projections arising from the site, program, and a building's method of construction, arguing:

> If the naturalism and graphic virtuosity of the architectural portrayals are too great, if they lack "open patches" where our imagination and curiosity about the reality of the drawing can penetrate the image, the portrayal itself becomes the object of our design, and our longing for the reality wanes because there is little or nothing in the representation that points to the intended reality beyond it.[4]

To resist the visual similitude of the drawing, Zumthor encourages architects to explore the construction of representations as a method to give voice to places that have not yet found physical presence (Figure 4.1). An architect who works in diverse media knows that different materials have different effects and affects that resist the draftsman's will to create desired forms. It is because of this confrontation with the construction of drawing and building that the material of one can create a resistance for the other. In the discussion that follows, this resistance between the drawn and constructed building will be explored, looking at how it can afford the architect a medium for drawing-out new designs.

THE FACTURE OF DRAWINGS AND BUILDINGS

Today, much of the practice of architecture is removed from the physical act of building and architects must employ different graphic means, including drawings to study and explore proposed constructions. For architects, though, the word "drawing" denotes a variety of visual imagery including construction documents, design drawings, analysis, details and sketches made on a computer or by hand with a pencil, pen, stick of charcoal, or crayon, to name only a few. Yet, drawing tools and their processes are not the same nor do they produce the same results. The hand moving a pencil across a drawing surface does not naturally produce a consistent line, but one that is varied, both thick or thin. This change in the quality of the mark is important, since different lines can produce different associations in the mind of the architect.

To consider a drawing in the same way as its construction is to consider it as a record of how it was factured. The word "facture" derives from the past participle of the Latin verb "*facio*," or "*facere*," which has the double meaning of both "to make" and "to do."[5] On a construction site, the craftsman utilizes different materials that require different methods of transforming and assembling them into architecture. Like buildings, drawings are also constructed of different materials that create different kinds of factures. For an architect every stroke of the pen or pencil, charcoal mark, smudge, erasure, and blur in the drawing has a signaling function and a meaning for the construction and experience of a proposed building. In this way the drawing becomes both a medium and a surrogate for the physicality of the building in absentia.

The ability for an architect to use the facture of drawings to study and explore the facture of buildings implies there is also an interconnection between our perception of building construction and our perception of drawing construction. In a discussion about the phenomenology of space, the French philosopher of the material imagination Gaston Bachelard observes how a visitor to a space may enter it but the space also enters and occupies the visitor.[6] Bachelard argues:

> Of course, thanks to the house, a great many of our memories are housed, and if the house is a bit elaborate, if it has a cellar and a garret, nooks and corridors, our memories have refuges that are all the more clearly delineated.[7]

Similarly, when drawing, the hand inscribing paper with ink, graphite, or pastel is in direct collaboration with the draftsman. Working in this way, the image of a proposed project arises simultaneously with the architect's imagination and the sketch mediated by the hand. In reference to painting, the French phenomenological philosopher Maurice Merleau-Ponty makes a comparable claim: "The painter 'takes his body with him,' says [Paul] Valéry. Indeed we cannot imagine how a mind could paint."[8] Certainly an architect also experiences a dependency upon their memories of matter and space in the conception of architecture. Thinking is not alienated from bodily experience but articulates, distils, and organizes it. It is this embodiment of lived reality that permits the transposition of our experiences and observations including the facture of drawings and the facture of buildings.

Although traditionally associated with its verbal and literary application in the construction of metaphors, our capacity to create cognitive associations between things arises directly from our perception of the world. As George Lakoff and Mark Johnson reveal in *Metaphors We Live By*, our being in the world and our body-centered mode of perception form the basis of cognitive associations in our conceptual system.[9] In a discussion about synaesthesia in particular, the neuroscientist Vilayanur Ramachandran notes that what "artists, poets and novelists all have in common, is their skill at forming metaphors, linking seemingly unrelated concepts in their brain."[10] For the philosopher Mark Johnson, "metaphor is perhaps the central means by which we project structure across categories to establish new connections and organizations of meaning."[11] In the hand of an architect, a metaphor can be more than a visual allusion but also a tool for solving the facture of space and material in a building, as, for example, Steven Holl had employed with the facture of instrument and sound in his Stretto House project. This use of a metaphor to create connections between the qualities of two things we experience implies that there can also be a similar mirroring of drawing and building. In his recently published "grimoire" of architectural drawing, the Italian architect and theorist Marco Frascari promotes a similar use of drawing that he argues originated as a mimesis of building construction, concluding:[12]

> Therefore, our conceptual system is generated by the architecture around us; we make buildings and they make us. Architecture is framed by embodied experience and embodied experience is framed by architecture, and this mirroring action is also embodied in the drawings.[13]

Nevertheless, physically drawn lines are not building lines and can hardly give one a true impression of the sensible experience at a future place. Rather, architectural drawings may be employed like metaphors in the critical transportation of sensory information from one modality to another.[14] For an architect who seeks inspiration from the facture of their drawings, the efficacy of a drawn facture therefore depends upon how well it relates to the facture of a building and vice versa.

MATERIAL AND RESISTANCE IN DRAWING-OUT CONSTRUCTIONS

Because different materials create different kinds of factures, different media cannot perform as universal signifiers of "stuff" but only those particular to their nature inviting different material reveries. In the hand of a draftsman a charcoal or pastel stick will resist representing the hard lines of plate glass in the same way an ink pen will frustrate the depiction of light diffusing into a deep space. For Anton Ehrenzweig, the creative development of architectural designs encourages precisely this loss of "conscious control" over the medium and preferences techniques that create "vague and open-ended" drawings "so as not to pre-empt the solution at too early a stage."[15] By releasing "conscious control" of a drawing or model to its facture, the architect is left to engage in a dialogue with its material processes. As a representation of architecture, each additional mark or element in the emerging graphic construction invites speculation.

The American and Spanish architects Louis Kahn and Enric Miralles were also deeply interested in the role that their drawing media could play in drawing-out their designs. Kahn's drawings frequently show a bold hand altering and reworking, emphasizing edges and boundaries with the tip of a charcoal stick. In his plan of the Kimbell Art Museum, immersed in the making, the use of charcoal resists Kahn's ability to erase or recompose his plans, forcing him to build up lines or parts to create emphasis relying on smudging-out and over-drawing. Conversely, Miralles's use of a crayon to sketch boundaries and spaces of the Mollet del Vallès Park and Civic Center explores the medium to both suggest built form and defy alteration (Figure 4.2). Without erasing, Miralles varies the pressure of the crayon across the page; sometimes light tentative lines suggest areas of possible form compared to more determined patches of tight zigzag strokes that fill seemingly random lozenge-like shapes. In a discussion about his design process, Zumthor places great importance upon the role that the facture of his drawings played as an aid for developing the design of his Vals Thermal Bathhouse, showing that the constructive imagination is rooted not in the form but in the resistance of material.

Completed in 1996, the Vals Thermal Bathhouse is famous for its use of long, thin horizontal slabs of local Valser quartzite stone to create a series of fifteen monolithic structures that house different temperature baths, showers, steam rooms, and resting spaces in a rich sensorial environment of sight, sound, and scent (Figure 4.3). At the beginning of the bathhouse design, Zumthor recalls how several sketches, representing what he called "boulders standing in the water," began to guide the successive iterations of the project's development.[16] These drawings that Zumthor referred to initially as "quarry sketches" or "block drawings," frequently appear in publications about the project and consist predominately of rectangular black-and-

4.2   Enric Miralles, Preliminary crayon plan sketch for Mollet del Vallès Park and Civic Center, Barcelona, Spain, 1992–1995.

4.3   Peter Zumthor, *Bathing Level of the Valser Therme*, 1996.

blue blocks created by dragging the shaft of a pastel crayon across pieces of tracing paper from either the top to bottom or left to right.[17] While the black pastel "blocks" are more densely fitted together at the top of the drawings, they are dispersed between blue smudges and black horizontal or vertical lines at the bottom. A careful comparison between the preliminary design and as-built drawings of the bathhouse reveal an underlying method to their construction: in almost every charcoal drawing created for the bathhouse, the blocks have been progressively layered from top to bottom in the same way that visitors were intended to "meander" through the structure from entrance to changing rooms and from the outdoor pools to the view of the valley below (Figure 4.4).[18] No eraser marks can be found in the drawing, indicating that the architect did not reconsider or rework the drawing on the page. Rather, each drawing in the set was executed and completed quickly, one following the next, exploring another draft of the architect's idea.

By dragging pastel sticks across the paper, Zumthor sought to develop a method of constructing the drawing that would help him both conceive and communicate the spatial and constructive implications of his proposed project. However, conventional methods of denoting architectural elements including walls, windows, stairs etc. are absent from Zumthor's early plan drawings. As Zumthor recalls:

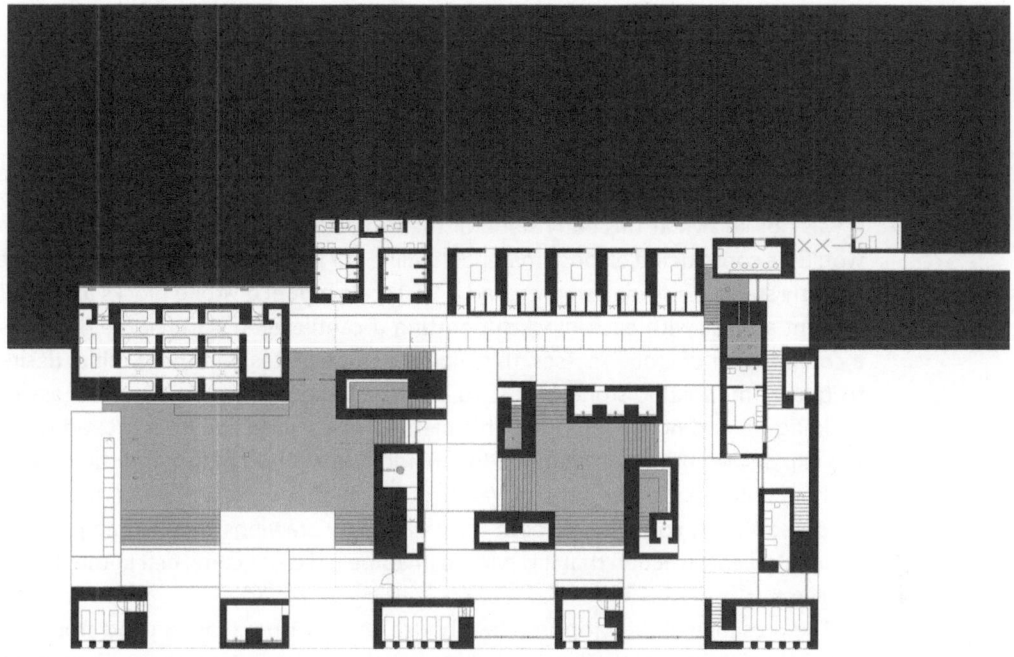

4.4   Peter Zumthor, *Plan of Bathing Level at the Thermal Bath House*, Vals, Switzerland, 1996.

4.5   Peter Zumthor, *Stone Model of Thermal Bath House,* Vals, Switzerland, 1990.

I remember feeling great freedom in pursuing issues of composition, working them out on the basis of these block studies, giving them shape in spontaneous drawings and trying to understand them by talking about them.[19]

Rather, the pastel sticks afforded Zumthor a method of facturing a drawing of the proposed project "as if it were a quarry, carving huge blocks out of it and adding others." While the image of a quarry occupied and fired Zumthor's imagination, it was not known at this early stage of the project design that the thermal baths would be constructed of loose material. "Unlike our block studies—they knew."[20] At an early stage Zumthor also prepared a 1:50 scale model of stone blocks arranged at right angles, with each block supporting a cantilevered slab (Figure 4.5).[21] As a composition of stone masses, this model demonstrates Zumthor's initial desire to quarry monolithic stone blocks to construct the bathhouse by a process of building up and hollowing out.[22] This method of construction later proved to be too expensive and was exchanged for an additive method of forming the masses by horizontally stacking slabs of Valser quartzite stone. Nevertheless, the stone model so closely compares to both the preliminary drawings and the image of the completed construction that it is hard to imagine it did not come first in the design process.

The presence of the proposed bathhouse within rather than beyond stones or smudges of pastel can be illuminated through a reexamination of Walter Benjamin's concept of the "aura" in a work of art. In his famous essay, "The Work of Art in the Age of Mechanical Reproduction," Benjamin distinguishes original from technically reproduced artworks through a concept of "aura."[23] For Benjamin, all original, unique art has an invisible aura that is not transcendental, but felt in the presence of its actual material, its making and aging over time. Yet, as Benjamin argues, when a work of art is re-produced, even as an exact copy, its aura is increasingly lost over time.[24] By searching for a way to use the quarried stones and pastel smudges lacking an aura, Zumthor combined them into new situations to impart a unique aura. When viewing Zumthor's drawings or models, one senses the presence of the aura that Benjamin describes. Furthermore, the use of the quarried stones rather than, for example, a drawn representation of them, allows the aura of the actual material to be present. In this way, Zumthor's bathhouse drawings and models appeal to the architect's material imagination because of their real material presences. The architect, in using pastel or stone, performs like Benjamin's magician in the act of interpreting actual present objects as architecture by being in a direct relation with them, while Benjamin's surgeon, who cuts into the object, loses a sense of a comprehensive relationship with it.[25] Because Zumthor's drawing and modeling method utilize the imaginative interpretation of the marks or configurations that the stone or pastel smudges afford rather than the representation of prior, fixed ideas imposed on material it sustains the architect's exploration of it as a work open to speculation. The French literary group *Ouvroir de la Littérature Potentielle* (workshop for potential literature), or simply Oulipo, were concerned with precisely this use of constraint in their own creative activities.

Founded in 1960, the Oulipo originated as a group of (predominately) French-speaking writers and mathematicians who sought to explore the creation of works using constrained writing techniques to trigger ideas and inspiration. The Oulipo have their beginnings in the science of 'Pataphysics (apostrophe intentional) invented by Alfred Jarry as "the science of imaginary solutions, which symbolically attributes the properties of objects, described by their virtuality, to their lineaments."[26] 'Pataphysics behaves as a philosophy of the *as if*, being simply the imaginary solution to the question *what if?* Nevertheless, as the founder of the Oulipo, Raymond Queneau, argues, to explore the *what if?* does not imply "blind obedience to every impulse."[27] The Italian writer Italo Calvino (who became a member of the Oulipo in 1973) concurs with Queneau, suggesting, "What Romantic terminology called genius or talent or inspiration or intuition is nothing other than finding the right road empirically."[28] This "right road" is dependent on the imposition of a structure or pattern, often based upon mathematical problems making Oulipians, as Queneau suggested, comparable to "rats who build the labyrinth from which they plan to escape."[29] By imposing his own constraint on the drawing of architectural plans without an eraser, Zumthor projects his own imagination into a labyrinth of pastel marks at the interstice between *wondering-at* and *wondering-about*.

The Romanian sculptor Constantin Brâncuşi argues that the importance which different media have in making art lies in their unique material natures:

> [y]ou cannot make what you want to make, but what the material permits you to make. You cannot make out of marble what you would make out of wood, or out of wood what you would make out of stone [...] That is, we must not try to make materials speak our language, we must go with them to the point where others will understand their language.[30]

If our sensible memories of architectural material and space are reservoirs for the design of places, it is the reality of building materials and structure whose properties an architect must use to develop an architecture that sets out from and returns to real things. Through his exclusion of the eraser, Zumthor aims to reverse the standard practice of design development from idea to plan to concrete object. As Zumthor argues:

> [t]he concrete, sensuous quality of our inner image helps us here. It helps us not to lose track of the concrete qualities of architecture. It helps us not to fall in love with the graphic quality of our drawings and to confuse it with real architectural quality.[31]

Yet working with concrete objects alone is not enough to generate the design for a project whose aim, as Brancusi has noted, can reside in their resistance to preconceived ideas.

As a form of resistance, the facture of drawing invites the architect to consider the corresponding making of a building to counter the arbitrary willfulness of a sometimes overly rampant designer's formal imagination. For Zumthor, drawing with the shaft of a pastel stick helped him to resist the "naturalism and graphic virtuosity of the architectural portrayal" to create "open patches" for the imagination to draw-out architectural ideas.[32] In these instances the facture of a plan drawing without an eraser has the same effect as the use of large unwieldy stone to create an architectural model since stone cannot be cut or manipulated into desired configurations of form. Rather, the concreteness of the stones used in the model determines particular configurations of architectural forms and spaces, instead of representing those predetermined by the architect. Conversely, the marks, accidents, and incomplete gestures of the drawings remain as "open patches" that engender the architect's imagination and invite creative speculation. In a description about his design process, Zumthor explains how the inner image of a proposed project emerges through a process of drawing:

> [w]ith the sudden emergence of an inner image, a new line in the drawing, the whole design changes and is newly formulated within a fraction of a second. It is as if a powerful drug were suddenly taking effect. Everything I knew before about the thing I am creating is flooded by a bright new light.[33]

As a draftsman, Zumthor may have begun by drawing black pastel masses, but as the marks begin to assert themselves, interacting with the guiding hand, they began to take on the character of the sought-for place—not as an illustration of an idea but as "an innate part of the work of creation, which ends with the constructed object."[34]

CONCLUSION

By directing the will of the architect, the use of factures as inspiration for new designs can invite an architect to think with the medium of architecture and not about it. Where the aura of an imagined place is in the facture of its materials, metaphor demonstrates how the facture of an architectural drawing can act as a medium for drawing-out the construction of architecture—one in which each additional mark on the drawing surface encourages the architect to speculate and imagine how they contribute to a proposed design emerging between material and the architect.

BIBLIOGRAPHY

Bachelard, Gaston, *The Poetics of Space* (Boston, MA: Beacon Press, 1994).

Bénabou, Marcel, "Rule and Constraint," in *Oulipo: A Primer of Potential Literature*, ed. and trans Warren F. Motte (1983; Lincoln, NE and London: University of Nebraska Press, 1986).

Benjamin, Walter, "The Work of Art in the Age of Mechanical Reproduction," in *Illuminations*, trans. H. Zohn (London: Pimlico, 1999).

Berger, John, *Berger on Drawing* (Cork: Occasional Press, 2nd edn., 2007).

Calvino, Italo, "Cybernetics and Ghosts," in *The Uses of Literature* (San Diego, CA: Harcourt Brace & Company, 1986).

Dal Co, Francesco and Mazzariol, Giuseppe, *Carlo Scarpa: The Complete Works* (Milan: Electa; New York: Rizzoli, 1984).

Eco, Umberto, *Experiences in Translation*, trans. A. McEwen (Toronto, ON: University of Toronto Press, 2001)

Ehrenzweig, Anton, *The Hidden Order of Art* (Berkeley, CA: University of California Press, 1971)

Frascari, Marco, *Eleven Exercises in the Art of Architectural Drawing: Slow Food for the Architect's Imagination* (New York, NY: Routledge, 2011).

Hauser, Sigrid, *Peter Zumthor Therme Vals*, ed. Peter Zumthor (Zurich: Verlag Scheidegger & Spiess, 2007).

Jarry, Alfred, "What is Pataphysics?," *Evergreen Review* 4, no. 13 (1963), pp. 131–51.

Johnson, Mark, *The Body in the Mind: The Bodily Basis of Meaning, Imagination and Reason* (Chicago, IL: University of Chicago Press, 1989).

Lakoff, George and Mark Johnson, *Metaphors We Live By* (Chicago, IL and London: University of Chicago Press, 1980).

Mallgrave, Harry Francis, *The Architect's Brain: Neuroscience, Creativity, and Architecture* (Chichester and Malden, MA: Wiley-Blackwell, 2011).

Merleau-Ponty, Maurice, *The Primacy of Perception: And Other Essays on Phenomenological Psychology, the Philosophy of Art, History and Politics*, ed. James M. Edie, trans. William Cobb (Evanston, IL: Northwestern University Press, 1964).

Motte, Warren F., Jr. (ed. and trans.), *Oulipo: A Primer of Potential Literature* (1983; Lincoln, NE and London: University of Nebraska Press, 1986).

Ramachandran, V.S., *A Brief Tour of Human Consciousness: From Imposter Poodles to Purple Numbers* (New York, NY: Pi Press, 2004).

Summers, David, *Real Spaces: World Art History and the Rise of Western Modernism* (London: Phaidon, 2003).

Zumthor, Peter, *Thinking Architecture* (Baden: Lars Müller Publishers, 1998).

NOTES

1   Maurice Merleau-Ponty, *The Primacy of Perception: And Other Essays on Phenomenological Psychology, the Philosophy of Art, History and Politics*, ed. James M. Edie, trans. William Cobb (Evanston, IL, 1964), p. 183.

2   James Elkins, excerpted from a letter to John Berger on January 29, 2004. Reproduced in *John Berger, Berger on Drawing* (Cork, Ireland, 2nd edn., 2007), p. 212.

3   Francesco Dal Co and Giuseppe Mazzariol, *Carlo Scarpa: The Complete Works* (Milan; New York, 1984), p. 242.

4   Peter Zumthor, *Thinking Architecture* (Baden, Switzerland, 1998), p. 13.

5   For this term, I owe a debt to Marco Frascari for introducing me to David Summers's discussion material and technique in painting. David Summers, *Real Spaces: World Art History and the Rise of Western Modernism* (London, 2003), p. 74.

6   Gaston Bachelard, *The Poetics of Space* (Boston, MA, 1994), p. 8.

7   Ibid., p. 8.

8   Merleau-Ponty, p. 162.

9   George Lakoff and Mark Johnson, *Metaphors We Live By* (Chicago, IL and London, 1980).

10   V.S. Ramachandran, *A Brief Tour of Human Consciousness: From Imposter Poodles to Purple Numbers* (New York, NY, 2004), p. 71 after Harry Francis Mallgrave, *The Architect's Brain: Neuroscience, Creativity, and Architecture* (Chichester and Malden, MA, 2011), p. 174, n 36.

11   Mark Johnson, *The Body in the Mind: The Bodily Basis of Meaning, Imagination and Reason* (Chicago, IL, 1989), p. 171 after Mallgrave, p. 171, n 47.

12   Marco Frascari, *Eleven Exercises in the Art of Architectural Drawing: Slow Food for the Architect's Imagination* (New York, NY, 2011), pp. 96–102.

13   Ibid., p. 6.

14   Umberto Eco, *Experiences in Translation*, trans. A. McEwen (Toronto, 2001), p. 74.

15   Anton Ehrenzweig, *The Hidden Order of Art* (Berkeley, CA, 1971), pp. 44–5.

16   Sigrid Hauser, *Peter Zumthor Therme Vals*, ed. Peter Zumthor (Zurich, 2007), p. 27.

17   Hauser describes the material and method of facturing the drawings in ibid., pp. 38, 58.

18   Ibid., p. 80.

19   Ibid., p. 38.

20   Ibid.

21   Peter Zumthor explains that he created the model during the 1:50 scale of design for the Thermal Bath House in Vals, Switzerland and later used it for an interview with the community to gain approval for the project construction. Later, during the development of the project, its form and construction changed. From Peter Zumthor, June 23, 2004, in an e-mail to the author.

22   Hauser, pp. 38, 42.

23   Walter Benjamin, "The Work of Art in the Age of Mechanical Reproduction," in *Illuminations*, trans. H. Zohn (London, 1999), pp. 211–44. See specifically, pp. 214–16.

24   Ibid.

25   Ibid., p. 227.

26   Alfred Jarry, "What is Pataphysics?," *Evergreen Review* 4, no. 13 (1963), p. 131.

27   After Marcel Bénabou, "Rule and Constraint," in *Oulipo: A Primer of Potential Literature*, ed. and trans. Warren F. Motte (1983; Lincoln, NE and London, 1986), p. 41.

28   Italo Calvino, "Cybernetics and Ghosts," in *The Uses of Literature* (San Diego, CA, 1986), p. 13.

29   Warren F. Motte, Jr. (ed. and trans.), *Oulipo: A Primer of Potential Literature* (1983; Lincoln, NE and London, 1986), p. 22.

30   Quoted by Dorothy Dudley, "Brancusi," *The Dial* 82 (February 1927): p. 124.

31   Zumthor, p. 59.

32   Ibid., p. 13.

33   Ibid., p. 20.

34   Ibid., p. 14.

5

Mythic Geology: Under the Surface at Palazzo del Te

Tracey Eve Winton

Rustication ain't pretty, but there's something about it gets under your skin.

Sebastiano Serlio felt it, and his treatises helped this curious, crude idiom spread, becoming far more popular in the sixteenth century than earlier Roman and Tuscan models of rustic simplicity. Ancient rustication—vernacular, not a classical order— featured coarse-hewn massive stonework, fortifications and cyclopean walls of close-fitten boulders: unwieldy, irregular and patinated, too huge for man to cut or set. Everyone knows the Cyclopes built them, a troglodyte race of uncivilized giants with one eye in the middle of their head. Marble is earthborn, like the giants. And so is terracotta brick. And stucco. And plaster.

How to recognize rustication, because it doesn't really have rules, as you might notice from reading Serlio carefully: It works by thickening or deepening the material surface that gives you the feeling of looking inside of raw matter. That impression is amplified by modeling in low relief, almost sculptural, maybe evoking reptile hide or tree bark or a goat's shaggy pelt, or imitating unworked natural stone with inclusions and impurities. Or a vernacular of living surfaces, following each stone variety's natural lines of cleavage. The nearly planar surface in smooth ashlar breaks up into objects and textures in the rustic through ridging or hammering the finish. The effect is unsystematic, even though structural masonry naturally follows level courses. Lines widen and mortar joints sink back deeply from the face of the stone, which is either left rough-hewn to look unfinished or laboriously worked into a natural, crude appearance. The uneven surfaces, projections and crevices work by seizing and holding shadow, giving the overall façade a captive darkness that enhances its feeling of raw material and integration with the mortal world of time and secret things. Rough stone catches light in a different way, showing its crystalline inner form, but it also casts micro-shadows on and across the surface, the texture of which deepens color and through the incorporation of darkness effects a feeling of a solid and heavy mass. Most typically we see rustication in the lower registers of a building, or at binding joints such as foundation lines and in quoins, and it's used in a way that amplifies the horizontal and lateral emphasis of the base. When used elsewhere, like portals and keystones, it transfers that sense of strength and stability.

Besides Rome, medieval Florence and environs offered precedents. Michelozzo built Palazzo Medici like a kind of geode—the exterior bristling with crude ashlars; intricate and delicate within. His façade renders in masonry a gradation of refinement from the lower storey's deliberate coarseness of bossy blocks at street level, to flatter, but still articulated on the *piano nobile*, and a smooth upper storey under the cornice. It's as if the solid ground itself is the material substrate out of which a massive palace organically rises, its rustication austere, noble and civic, an idiom of the public realm.

When rusticated, architecture's flesh is conveyed by a vulgar treatment of its material, which deepens the liminal space of the surface and emphasizes separate stones in the assemblage. We experience our immediate relation to architectonic objects through a sense of corporeality, a near-erotic sympathy with fleshed-out objects and elements, a visceral and not logical sense of recognition in them of some of our own qualities. Where we encounter it in tactile range, rustic stone's corporeal protrusions, at a scale resonant with our own bodies and limbs, create an archaic sympathy of living matter, connecting flesh and blood with our mineral kin.

Our feeling for architecture responds to material beauty as well as the form in which it gets shaped. By material beauty I mean "form already visible in matter," like wood-grain, the threads in a textile, the crystalline structure in ore, or veining in marble. We don't normally recognize or acknowledge the inward organization of material as form. It tends to fly under the radar of everyday perception, although jewel-cutters and woodworkers know it well. This inward order responds to an outward order brought by cutting or carving, or wear from the elements, a dialogue between the "outward" laws of craft and the arts, and the hermetic "inward" laws of Nature. When the two are proportionate you could call it a conversation, or a battle when they don't align, and each asserts violently against the other.

In architecture, form describes the external lineament or boundary defining an element in space, the horizon of our awareness in physical things. Rustication is an architectural attempt to render visible matter's secret life. The root comes from the Latin for the coarser things of the countryside and its natural settings, *rus* in contrast to *urbs*, the civil society of the city. Going back further, the Proto-Indo-European root *reуǝ-* refers to an open, undefined expanse.[1] Turning to its cognates, there's imagery of chewing, intestines, digging or tearing out, and ruins: processes that convert definite form to the formless. If we consider the design process as intent to define form, the rustic offers back to the material denatured by imposed geometries a chance to recuperate some of its vital character by exposing the raw interior.

Serlio mentions "that most beautiful palace called the Te," a building then very recent, in his discourse on how the Ancients used to mix Rustic with the other orders:

> by this representing, part a work of Nature, and part a work of artifice: such that the columns with straps of rustic stone, and also the architrave, and frieze interrupted by voussoirs were demonstrating the work of Nature, while the capitals and parts of the columns and thus the cornice with the frontispiece were representing the works made by the hand of man: which combination, in my opinion, is very pleasing to the eye and is in itself a great fortitude.[2]

5.1   Frontispiece of *Il Terzo Libro*, Sebastiano Serlio. Venice: Francesco Marcolini, 1544.

So let's skip antiquity, for now, and fast-forward to Isabella d'Este's Mantua.

In 1506, the painter Andrea Mantegna died, bereaving the city of its artistic genius. Isabella's son Federigo Gonzaga had inherited her love of arts and culture, and in 1520 he sent Baldassare Castiglione, author of *The Courtier*, to entice from Rome Raphael's accomplished protégé Giulio Romano. Federico was the same age as Giulio, and as a boy had been a political hostage in Rome, where Raphael painted his portrait. Around 1524, the young marquis commissioned the architect to design a villa on the rustic foundations of a farm building at the marshy south end of Mantua, on the "island" of Te, where the family stabled their famous horses.

5.2 Courtyard façade (west face) of Palazzo del Te, Mantova, by Giulio Romano, 1535.

Giulio was an antiquarian with classical expertise. He grew up in Rome at the foot of the Capitoline Hill, and was intimate with the fragmentary, half-buried remains of antiquity. This project gave him a chance to find his own voice and establish his position on architecture in relation to history. Federico had something in mind too: a stylish villa for romantic escapades, away from his mistress's husband and his own furious mother.

If you don't know what Palazzo del Te looks like, look it up; it will keep you busy. Giorgio Vasari characterized it as a quadrangular building of Roman grandeur around an open-air courtyard into which debouch four entrances in the form of a cross. The primary entrance passes into a large loggia, which opens into the garden while two others lead into the four apartments adorned with stucco-work and paintings. The outside has two façades dressed with rusticated surfaces, each with an entrance. One features a triple portico facing north towards the city, across the canal, at the receiving end of the "Prince's Route" from the ducal palace. It is thus "civic," and its orientation to the civilizing arts of the city is reinforced on the inner face, in a loggia dedicated to the Muses. The other portal faces west, and leads through the Great Vestibule. Before Giulio's renovation, planted allées had elongated this axis east and west of the building. These two entrances into the internal *cortile* imply a *cardo* and *decumanus*, the first rudiments of form sketched out in space, and a symbolic crossing of Art and Nature. The inner four faces, each distinctive, are dressed in similarly complex style amalgamating the Doric order with rustic masonry. These façades look like nothing ever before.

Everything's more complicated than it first appears. Spoiler: most of the villa's powerful-looking stonework is actually stucco troweled over load-bearing brick masonry. Which, however, has been ingeniously contrived for that purpose, a cleverly structured skeleton waiting for flesh, on top of the reused foundations. Mantua wasn't a place with living rock or quarries from which to excavate material for hewn or carved stone, so Necessity brought on Invention, which, judging by both Vasari's and Serlio's writings, seems to be the most desirable trait of an architect in this period.

We might think of plaster or stucco as a white, unifying surface, but it's really a mixture of water with a binder like lime, gypsum, or animal glue, using sand as an aggregate, like concrete. It's a manmade species of mud: a very archaic building material. Not unrelated to clay brick, in the sense you can really get your hands into it while it's still wet. Anyhow, it was the stucco reliefs found in Nero's submerged Domus Aurea in 1480 that ignited an interest in *grottesche* or "grotesques" as they were called, when the buried palace was misconstrued as an underground grotto, an old myth materialized in built form. The term "grotesque" is cognate with *grotta*, a grotto, and *crypta*, a vault or cavern, denoting a hollow in the earth or something secreted underground: a rich symbol for the distinct experience of interiority, and archaic esoteric wisdom. In 1498 when Mantegna was still alive, and possibly involved, Isabella d'Este writes of a "Grotta" she had built under her *studiolo*, complete by a decade later, the first of two.[3] This new fashion in architecture had been introduced to practice by none other than Giulio Romano's beloved master, Raphael, one of the first artists to climb down into the underground complex after it was discovered, and it was Raphael who named the fantastical style "*grotteschi*."

The rough manifestation of building materials has its roots in the mythic topos of the grotto. There's also something of the original "grotesque" in Giulio's approach of concealing an armature under a fictive, crudely articulated surface. On the Palazzo's north face, the elements look greatest and most monolithic as they hug the earth, lightening in appearance as they rise. From the architect's perspective, the huge blocks of crudely hewn stone seem to structure the wall. As you go deeper, you notice the inset pilasters, a conflicting system, without respecting modular rhythm, replacing regular column bays with incremental compressions towards the edges, and expansions at the center, that impart a sense of movement and flux. Besides stone-courses of blocks, the masonry appears also in arches. Pilasters project enough to sever the set-back lintel course with the Greek key carving, and the keystones of the triple portal rupture it also. Giulio's mixes are unorthodox; his spacings inconsistent; his structural systems clearly fictitious. Window openings are irregular, but not enough that you notice at first. Blocks burst forward and retract, smooth and rustic finishes combine. Triglyphs and blind windows run in sequences independently of other elements. If you enter the north portico, the rustic finish making up the rectangular pillars turns out to be on the front only; the sides are not only smooth but monolithic, without inscribed mortar joints. The question on the table is: what is this *stuff*?

In Florence, Michelozzo's simple sequence of blocks shows unformed matter caught in the act of evolving towards geometrical form, or divine idea, metaphorically climbing a great chain of Being from the stone's chthonic birthplace towards Olympic heights. A century later Michelangelo, steeped in the same Neoplatonic philosophy, left his famous "slaves" in the state of *non-finito*, figures caught in a struggle to free themselves from the marble block or collapsing back into it. In similar ways they recognized matter's mythic geology, its continuous cycles of transmutation. Michelangelo felt a protean spirit in the stone. These sonnet lines that he wrote around 1540 declare material's primacy over the vicissitudes of form, recalling Aristotle's image of the statue of Hermes latent in a block of wood: "Even the great artist can have no conception that isn't circumscribed within the superfluous marble of a single block, and that only the hand that obeys the intellect may realize." Since the artist (they say) slept with Dante's epic beside him, it's noteworthy that his word *superchio*—superfluous, excess—is a descriptive of the rocky Inferno, the unfathomable abyss, as the pilgrim begins his plunge into the architecture of the earth.

The most beautiful space in the palace is the Great Vestibule at the west entrance, a material *parti* of the whole complex. There Giulio shows his mastery of a Roman mode of invention, hoisting on four robust columns of rustic finish (think satyrs' shaggy legs) the barrel vault of the Basilica of Constantine, the most striking ruin in the Roman Forum. Its distinctive coffering is delineated in the grotesque style by egg-and-dart moldings and astragals that suggest vertebrae, pebbles and shells. As you enter through the archway under its exaggerated keystone and heavy rustication the cool vestibule becomes a dark optical canal framing in sequence the bright opening to the far portal into the Loggia of David, the fishponds and the garden. This threshold zone, striking enough to make you stop and pause, transversally re-orients and delays you, ceremonially expanding the moment of transition and effectively deepening the material plane of the building so that you find yourself right in the heart of matter. Symbolically and effectively, you are in an architecture of chorous fluidity demonstrating a chiasm between Nature and Antiquity.

The Vestibule is a frame to appropriately dispose you to grasp the iconographic program within. Here Nature greets you undisguised as a dynamic if perilous force in the form of a column, a proxy human body. Grottoes were thresholds linking the upper and lower worlds, portals into the *prima materia* of the ground. As entrances to the underworld, they are symbolically places of living, growing matter, and painted grotesques, following the model of the Domus Aurea, showed monstrous creations that amalgamated animal, vegetable and mineral elements.

The well-known Renaissance garden grottoes were only created decades later, including one in Palazzo del Te's secret garden: a marvel of encrustation and color, as if the rock were sprouting with life, overspilling its boundaries. Architects followed descriptives in Leon Battista Alberti's treatise on building, reflected in their use of unstandardized materials that were collected and hardly processed, if at all, like coral, mollusk shells, river pebbles, or porous volcanic rock, continuing to evidence their close, unsevered connection to the fertile matrix of the earth. That by architecture's geometric standards of order these materials were "formless" was intentional. Using rough, uneven material as ornament symbolizing Nature, artificial grottoes forged the atmosphere of transformation by composing heterogeneous pieces into surfaces and projections that revealed emerging but incomplete geometries forging spatial relations, cosmological diagrams symbolizing order. In his *Paradiso*, Dante had offered this poignant image: Nature is an artist who knows his craft well, yet works with a trembling hand.[4] Grottoes in this way represented rudimentary, natural precursors to architecture, as Vitruvius recounts in his second book of *De Architectura*, where prior to building "men of old were born like the wild beasts, in woods, caves, and groves, and lived on savage fare."[5] Instability propels the imagination to anchor itself in myths of origin, the primordial and the elemental. When Saturn ruled the Golden Age, the rustic and the raw meant carnal pleasures, and Arcadia's pastoral idyll of satyrs and nymphs was a home in Nature. Then Jupiter banished his father and instituting divisions in space and time begot his Silver Age, with which came war and the foundations of architectural order. Now men had to build.

The Palazzo del Te had two ideal inhabitants: the lover and the architect, and in it each had his proper place. Two special frescoed rooms symmetrically arranged about the east–west "Nature" axis demonstrated opposing cosmic forces, love and strife: *eros* being the power of linking and joining, while *machia* is the principle of separation and fragmentation. To the north is the Room of Amor and Psyche, dedicated to scenes of love, decorated with nymphly beauty and Dionysian satyrs. To the south, the Room of the Giants, with its violent imagery of *il mondo sotterraneo*, what Giovanni Pico della Mirandola (philosopher of protean human reality) called the world of matter. If the Psyche room celebrated the body's voluptuous pleasures, the Giants room is manifestly about architecture.

5.3   Gigantomachy, Virgil Solis, p. 6 in Ovid, *Posthii germershemii Tetratischa in Ovidii Metamorphosis libri XV*, Frankfurt: G. Corvinum, S. Feyerabendt and W. Galli 1569. Bibliothèque nationale de France.

The Roman poet Ovid links Jupiter's Silver Age to the invention and first practice of architecture, supplanting man's primitive shelters under boughs and in caves. It is in the Iron Age, following Bronze, that the battle of the gods and giants takes place. "Giants piled hill on Mountain to make a stair that reached the skies," says Ovid.[6] This is the room with the double foundations and the double vault into which Giulio and visitors like Vasari poured their interest. Its unbroken fresco shows the gods and giants battling for supremacy, a well-established mythic tale of the forces of chaos and order. In a nutshell: When the king of the gods ruled from high on Mount Olympus, the giant Alcyoneus led a revolt against his citadel. The giants, earthborn creatures, attempted to storm Jupiter's throne and overthrow his rule. These primeval autochthons, brutal and hideous, have begun to climb towards the heavens, when Jupiter strikes them down with thunderbolts and lightning, crushing the revolt. Buried alive in the rocks, their raw brutishness pulses in pain.

> In the lower part, that is, on the walls that stand upright, underneath where the curve of the vaulting ends, are the Giants, some of whom, those below Jove, have upon their backs mountains and immense rocks which that they are supporting with their powerful shoulders, in order to pile them up and climb into the heavens, while their ruin is already underway, for Jove is thundering and all of the heavens burning with indignation; and it appears not only that the Gods are dismayed by the arrogant audacity of the Giants, upon whom they are hurling mountains, but that the whole world has been turned upside down and as if at its final end. In this part Giulio painted Briareus in a dark cavern, nearly covered with massive chunks of mountain, and the other Giants all crushed and some dead beneath the ruins of the mountains. Besides this, through a great hole in the obscurity of a grotto, which reveals a distant landscape rendered with good judgment, many Giants may be seen fleeing, all struck down by the thunderbolts of Jove, and, as it were, on the point of being overwhelmed at that moment by the ruins of the mountain, like the others. In another part Giulio fashioned other Giants, upon whom are falling temples, columns, and other pieces of buildings, making a vast slaughter and havoc of those proud beings. And in this part, among those falling fragments of buildings, stands the fireplace of the room, which, when there is a fire in it, makes it look like the Giants are burning, for there is painted Pluto, fleeing towards the center with his chariot drawn by parched horses, and accompanied by the infernal Furies; and thus Giulio, not straying from the subject of the story with this invention of the fire, made the fireplace a most beautiful ornament. In this work, moreover, to make it even more horrifying and terrible, Giulio represented the Giants, of enormous and fantastical stature, falling to the earth, smited in various ways by flashes of lightning and thunderbolts; some in the foreground and others in the background, this one dead, the other wounded, and still others buried by mountains and the debris of the buildings. Wherefore let no one ever think to see any work of the brush more horrible and terrifying than this, nor more naturalistic; and anyone who enters that room, on seeing the windows, the doors, and other thus-made things wrenched awry and on the very brink of collapse, with the mountains and buildings hurtling down, cannot but fear that everything will fall upon him ...[7]

Vasari describes the terrifying drama swirling around the visitor as fully in the round.

And what is most marvellous in the work is to see that as a whole the painting has neither beginning nor end, but is seamlessly joined and connected together, without any divisions or ornamental partitions, that the things which that are near the casements look very large, and those in the distance, where there is countryside, go on endlessly receding; whence that room, which is no more than fifteen braccia long, has the appearance of open country. Moreover, the floor being paved with small round stones mounted on edge, and the lower part of the upright walls being painted with similar stones, there is no abrupt junction to be seen, which makes that surface seem like a vast expanse …[8]

You are standing in the heart of the mountain. The uneven floor unbalances and slows natural movement. Far above you can see the peak. Jupiter's throne sits in a circular Ionic temple that pays homage both to Mantegna's oculus in the Camera degli Sposi, and beyond that to the coffered dome of Hadrian's Pantheon. On the throne perches an eagle, Jupiter's bird, like Mount Olympus itself, a Gonzaga emblem. Around their thunderbolt-wielding ruler gather the startled Olympians, occupying the upper part of the room rounded by the vaulting. In three directions you are surrounded by living rock fracturing into gigantic boulders. Trapped by an avalanche of rocks the fallen giants are crushed and mutilated, screaming in pain, and bleeding. Among them are Cyclopes, the muscular one-eyed workmen of Vulcan's forge. This is a room, quoth Vasari, built on the principles of a furnace, with all the constructed details rusticated: doors, window surrounds, fireplace. You are also inside Mount Etna, and the opening above is the volcano's crater. To the east, fires are raging. Originally, between the windows, a real fireplace merged with the hearth of the underworld. Rough stone blocks are interspersed with the protruding body parts and disembodied limbs: an arm, a hand, a knee, a head. Like Michelangelo's *Slaves*, the colossal figures struggle to lift the tumbling rocks off themselves, but their bodies have become integrated with the dense masses pressing down on them. Between shapeless lumps of stone and flesh tree trunks are caught. Gaping holes have opened in two flanks of the mountain and through these the damage spreads in waves through the elemental forces of nature.

To emphasize by contrast the rustic character of the cyclopean boulders that form the architecture of the mountain, Giulio renders on the north wall, at your entrance from the Loggia of David and thus closest to the Psyche room, a series of architectural elements of Roman refinement, cracking, breaking apart. Here real architecture collapses: a Hadrianic structure with Corinthian capitals of antique veined marble and detailed marble cornices, columns of heavy stone, ionic capitals, ornamental details very finely sculpted in white marble, and entire walls of brick. The disintegration is so advanced that it is inconceivable to reconstruct how it looked as a whole. While the imagery looks nothing like the villa's façades, it also mixes arcuated and trabeated structures, masonry courses and columns, and naturally evokes the ruins of Rome. You are constrained to the lower storey of the room, subject to earthquake and rockslide, in the subterranean zone of the giants.

At the spring points of the vaulting and thus at the level where the chaos of falling rock and earth touches the heavens, four personified Winds blast powerful storm-gusts and horns of war. Though this is no charming garden grotto, the Winds represent *spiritus*, and are also a clue to the special acoustic properties of the room

5.4 Room of
the Giants, Giulio
Romano, Rinaldo
Mantovano, and
assistants. Vaulted
ceiling, south and
west walls.

as an echo chamber in which all sounds are magnified into clamor. My feeling is that they are used, as Lucretius did in numerous passages in *On the Nature of Things*, to translate the Gigantomachy's chthonic forces from mythical to elemental, referring to a force trapped within matter and seeking to escape or express itself.[9] Above the level of massive stones, a layer of clouds in which the Winds are embedded forms the background for the Olympian legion. The shaping of the clouds translates the rustic shapes of the stone into an airy, insubstantial medium. The obscuring masses of grey clouds render an impenetrable barrier between the worlds and this miasma has a rhetorical meaning as a zone of complete formlessness which that functions to provide an ultimate referent to the quasi-formless character of the more-than-rusticated rock clusters that sustain the shell of the mountain in which we are standing. The clouds are morphologically related to the massive boulders which that they resemble, but firmly outside of architectural language. Still, once you look at the relation of the clouds and the boulders, some of the boulders seem to be loosening themselves from the bond of gravity in order to float upward, rather than tumbling down—much like the rusticated keystones in the courtyard, breaking the pediments above as they rise.

In Plato's dialogue *The Sophist*, the Eleatic Stranger presents the Gigantomachy as a battle staged over whether Being is bodily and material, or incorporeal and ideal, pitting the Giants versus the "friends of Ideas" or the Olympians of the myth.[10] The "gods" people break the bodies of the "giants" into bits, and call them a process of coming-to-be: Becoming rather than Being. The contrasts of Plato's allegory are apparent, where the Giants' colossal bodies are anatomized into limbs and members, to be compared by the visitor with a corresponding dismemberment of the architecture pictured on the north wall. Keep in mind the analogy to the architectural art of all the façades, though not precisely a post-Babel surrender to idiom. One wonders whether for Giulio the debate shifts from Being to Architecture, whether in essence it's a bodily or non-bodily thing, brought on by the classical orders' ideal forms (the divinities) in contrast with the decayed reality of the ruins in Rome (the Giants), and the question of how those two worlds connect. The fresco of the Giants shows a complex descent of matter, and describes the reduction of a whole into partial elements and fragments, whose fractured liminal regions lack the precision lineaments of architectural jointing, and thus the impossibility of crafted detailing. Just as grottoes with geometrical lines and relations starting to appear rising into visibility in the arrangement of coarsely shaped natural materials echo the transition from the undivided harmonious peace of the artless Golden Age to the Silver Age with its spatial divisions and architectonic forms. Partibility or disunity comes from discord, *machia*, and this is its space.

Plato's outcome, like Giulio's architecture, lies somewhere in-between, as between Being and Becoming lies *chora*, Plato's space of appearance. Attention

5.5   Room of the Giants, Giulio Romano, Rinaldo Mantovano, and assistants. North wall.

to Giulio's conception of space in relation to materialized form shows a pattern. We know that the Psyche room should have featured a statue of Venus holding the center. The great vestibule features pilasters around the edges, and fully disengaged columns in the center. These are in fact the only free-standing columns in the complex, the most bodily of all the forms materially rendered. The interior court shows a corporeal character to the architectural elements, the bodying-out of the round columns in particular, that contrasts with the comparatively flat inset pilasters of the exterior. This courtyard should have featured a labyrinth (another Gonzaga symbol) within it as planned in the drawings: a tactile field which that traces the horizontal plane of the material world and the earth and marks the entrance to the underworld.

To consider the Room of the Giants in light of the space of appearance, of the materialization of form, let's walk back through the city to the architect's model and reference, Mantegna's Camera degli Sposi (1467–74) in the Castello San Giorgio. Here, two figuratively painted walls are lighted by real windows, while the two other shadowy walls are painted with heavy false curtains. On one wall the city of Rome, an architectural fantasy of monumental structures, is surrounded by landscape in which men are extracting building stone from quarries deep in the earth. On the other the Gonzaga court preside like Olympic deities in a life-size portrait. The walls and golden vault are linked by frescoes depicting Ovid's myths, thematizing metamorphosis. Mantegna, master of perspective, creates an effect of solidity and stability in the room contrasted with non-solid things playing against this: draped fabrics, voids, and unstable surfaces. He constructed the volumetric space of this "painted room' with precise geometry, like the painted pilasters that seem to support the vaulting. Built material objects support the pictorial illusions: the frame around the fireplace supports the carpet that serves as a surface for the fictive staircase on which several court figures stand, one "in front of" a painted pilaster, and thus in "your" space. At the center of the vault above you, the ceiling is dematerialized by a trompe l'oeil oculus constructed with concentric circles. This giant cyclopean eye, a "hieroglyph" for Alberti's perspective, is open to the heavens, turning the enclosed room inside out, like a courtyard. From it, a troupe of foreshortened figures peers down; two of them are balancing on a rod a planter of wild quinces, symbols of marriage and fertility, about to tip it over on whoever is below.

The room's unified surfaces open a discourse about real space. The fruits poised to fall on you make you realize how your body takes primacy and focus by occupying the center of the imaginary space. Like the Room of the Giants, this artistry synthesizes real, lived presence and imaginary or historical space. In Book I of his treatise Della Pittura, Alberti describes the istoria painting as like an "open window"; perspective shows imaginary space beyond the wall surface.[11] Mantegna inverted his formula, and brought the imaginary depth to the inside, in front of the wall. This is why he painted false curtains on the third and fourth walls: to complete in detail the artificial or imaginary space in which you are standing. He also painted draperies on the pictorial walls, but drawn back to reveal the Gonzaga dynasty to you, or to subject your momentary, vulnerable presence to their stable, eternal gaze.

Perspectiva, a Latin word, literally means "looking through." Mantegna's closed curtains obliterate any picture-window transparency the wall might present, or perspectively conceived depth. Their "double-negation' restores to the wall its quality as an articulated plane, a sophisticated spatial thesis about a room, constructing the place of experience. Instead of the perspectival dematerialization of real architecture that opens a picture-window into imaginary space, Mantegna's heavy curtains reclose that aperture in order to affirm the material quality of the architectural stucco surface, but leave it ambiguously pliable and flowing with ornamental drapery.

In contrast to the Camera degli Sposi, the Room of the Giants uses a palpable screen of crude, crumbling rock and breaking architecture to rematerialize the space of the "picture-window;" its painted geological "structure" has a dense, solid thickness. This is not Alberti's infrathin picture- window that cleanly separates real and imaginary space from each other as if by a sheet of glass, nor Mantegna's delicate pilasters and patterned curtains. The two rooms, the noble and the rustic, run full of parallels and enjoy a subterranean conversation about space as the arena in which matter transforms. Giulio's last "postcard to Mantegna" in the Room of the Giants is his "reclosing" the oculus open to the sky with the dome of the Ionic temple.

The courtyard of Palazzo del Te reads like an outdoor room with its coherent revetment, staging the same forces and conflicts, materialized in the architectural surface, within which imaginary space opens up as tactile, not just visual. The gods are played by the Doric order with its austere masculine precision; the giants by the brawny Rustic. Each rough block, each deepening of the plane, is a victory for the Earthborn,; each smooth element stabilizing the surface, clarifying form, one for Jupiter. Every visible rhythm and symmetry belongs to the Olympians, every rupture of Vitruvian order, to the giants. The rising keystones, splitting apart the pediments as they press upwards, those are chthonic forces forcing their way into the heavens. The dropping triglyphs, centered on each bay of the west courtyard façade, taking chunks of the trabeation down with them: the thunderbolts hurled from above. The building's faces are the Gigantomachy: neither a fresco nor a sculptural frieze, but an architecture staging the battle of the powers of Form and Matter. If we move up a scale, we see the same pattern: the architectural skin of the façade is "at war" with an expansive pressure from inside the depths of space in the suites. The classical "rules" break on the exterior to accommodate idiosyncrasies in the rooms behind; real space transgresses the limits and overspills the boundaries of imposed geometry. The architect documents the mythic struggle to keep living Nature contained.

The rudimentary act of building divides us from Nature, which remains outside. The grotesque, the rustic, challenges this clean separation, asserting that Nature is always latent within. Threshold conditions proliferate and thicken the transition from one space to another, from one condition to another. The further into Palazzo Te you go, the greater the inside surpasses the outside in scale. The surplus space of the interior increases until finally in the Room of the Giants the entire phenomenal universe forcibly compressed with the world of divine ideas is captive in a single high-pressure chamber. The palpable obscenity of the rusticated, a quality of surplus and superabundance, has been brought inside, where you are standing: inside the frescoed room, inside the architect's mind, inside the mass of material, watching things actively move and change as the battle plays out. As only a few decades before Marsilio Ficino enquired in his *Platonic Theology*, "For after all what is human art? It is a sort of Nature handling matter from the outside. And what is Nature? It is art moulding matter from within, as though the carpenter were within the wood."[12]

At Palazzo del Te, Giulio Romano turned Roman brick-faced concrete inside-out, and questioned how form and material are related in architecture, a problem he may have first encountered in his childhood among Rome's ancient ruins and

5.6   The Great Vestibule, western entrance to the Palazzo del Te.

"grottoes." He inquired into the liminal zone between what is visible to us as form and the depth of "formless" matter lying below the horizon of vision, metaphorically underground, and the mysterious passage back and forth between them. On the one hand is what we now call entropy, destabilization and dissolution from defined wholes into fragments towards the grotesque. On the other, the metamorphosis of the rustic, raw materiality of Nature towards the geometrical or figural precision of human art and the stability of form: the province of the architect. Thus Nature and Art complete each other in a simultaneous cycle of corruption and generation.

Giulio's project is a manifesto concerned with this question, to which the Room of the Giants is a key. The Great Vestibule suggests that he saw the grotesque functioning as a material space of appearance. The four swollen, shaggy rustic columns through which you pass are framed with finely worked, smoothly finished bases and capitals to emphasize by contrast the deliberate expression of the inward material character of the stone, what a later writer, Giovanni Cadioli, called "the noblest rusticity." They are the only free standing columns of any order in the villa, and the only structural elements of solid marble. This establishes them at the pinnacle of the hierarchy of architectonic elements. They are four, the number of the material world, and they represent Nature at the heart of the architecture not as a finished object but as an ever-present matrix of arcane operations, *natura naturans*, which, outside of the poets, we know only through phenomena.

If you don't believe my story, look at the figural iconography sculpted into the façades, besides the obligatory Gonzaga *imprese*, and the instruments of war which that *per forza* refer to the Gigantomachy. Ask yourself why so many Doric metopes in the frieze, as well as the bridge that joins the palace and the garden, bear relief-sculptures of grotesque satyr masks. The satyr, mixing manly "Doric" and goatish

5.7   Satyr face between rusticated voussoirs, Palazzo del Te.

"Rustic," a creature that throngs the Amor and Psyche frescoes, symbolizes the rustic, feral and procreative in human nature, along with the imagination's fertile powers to bring things into being from a space within the turbulent wilderness of matter.

Like you standing bodily in the Camera degli Sposi, the Room of the Giants, or the courtyard, these most corporeal of architectural elements suggests a flicker of goatish gesture breathing in the dark caverns of history, a fluxuous material presence with innate transformative powers that can turn the world inside out, and leave traces of its raw nature on the exterior surface.

BIBLIOGRAPHY

Alberti, Leon Battista, *Della pittura e della statua di Leonbatista Alberti* (Milan, 1804).

Dante Alighieri, *La Divina Commedia, Tom. III: Paradiso*, ed. B. Lombardi (Rome: de Romanis, 1822).

Fejfer, Jane, Fischer-Hansen, Tobias and Rathje, Annette (eds), *The Rediscovery of Antiquity: The Role of the Artist* (Copenhagen: Museum Tusculanum Press, 2003).

Ficino, Marsilio, *The Platonic Theology*, Volume I, trans. M.J.B. Allen, ed. J. Hankins (Cambridge: The I Tatti Renaissance Library, 2001).

Lucretius, *On the Nature of Things*, trans. M.F. Smith (Indianapolis, IN: Hackett Pub Co., 2001).

Ovid, *The Metamorphoses*, trans. H. Gregory (London, 1959).

Plato, *Volume II Theaetetus | Sophist*, trans. H.N. Fowler (New York, NY: Putnam, 1921).

Pokorny, Julius, *Indogermanisches Etymologisches Wörterbuch* (Bern: Francke, 1959).

Serlio, Sebastiano, *Tutte l'Opere d'Architettura di Sebastiano Serlio Bolognese, Dove si trattano in disegno, quelle cose, che sono più necessarie all'Architetto* (Venice: Francesco de' Franceschi, 1584).

Vasari, Giorgio, *Le vite de' piú eccellenti pittori scultori ed architettori, Tomo V.* (Florence: Sansoni, 1880).

Vitruvius, *The Ten Books on Architecture*, trans. M.H. Morgan (Cambridge, MA: Harvard University Press, 1914).

NOTES

1   Julius Pokorny, *Indogermanisches Etymologisches Wörterbuch* (Bern, 1959).

2   "E stato parer de gli antichi Romani mescolar col rustico non pur il Dorico: ma il Ionico, e'l Corinthio ancora; ilperche non sarà errore se d'una sola maniera si fara una mescolanza, rappresentando in questa, parte opera di natura, & parte opera di artefice: percioche le colonne fasciate dalle pietre rustiche, & anco l'architrave, & fregio interrotti dalli conij dimostrano opera di natura, ma i capitelli & parte delle colonne, & cosi la cornice col frontispicio rappresentano opera di mano: laqual mistura, per mio aviso, è molto grata all'occhio, & rappresenta in se gran fortezza." In *Tutte l'Opere d'Architettura di Sebastiano Serlio Bolognese, Dove si trattano in disegno, quelle cose, che sono più necessarie all'Architetto* (Venice, 1584), the page between 133 and 134.

3   Beth Cohen, "Mantua, Mantegna and Rome: The Grotte of Isabella d'Este Reconsidered," in Jane Fejfer, Tobias Fischer-Hansen and Annette Rathje (eds), *The Rediscovery of Antiquity: The Role of the Artist* (Copenhagen, 2003), pp. 323–70.

4   Paradiso XIII, 76–78, translation by the author. "[M]a la natura la dà sempre scema, similemente operando a l'artista ch'a l'abito de l'arte ha man che trema," in B. Lombardi (ed.), Dante Alighieri, *La Divina Commedia, Tom. III: Paradiso* (Rome, 1822), pp. 194–5.

5   Vitruvius, *The Ten Books on Architecture*, trans. M.H. Morgan (Cambridge, MA, 1914), Chapter I: The Origin of the Dwelling House, p. 38.

6   Ovid, *The Metamorphoses*, Book I, trans. H. Gregory (London, 1959), p. 35.

7   "Giulio Romano," in Giorgio Vasari, *Le vite de' piú eccellenti pittori scultori ed architettori, Tomo V.* (Florence, 1880), pp. 535–44. Translation by the author.

8   Ibid., pp. 543–4.

9   Lucretius, *On the Nature of Things*, trans. M. Smith. (Indianapolis, 2001). For example, in Book VI, Lucretius treats the causes of phenomena connected to this mythical topos, like clouds and the impact of the winds on bolts of thunder and lightning (96–422) and on earthquakes (535–607): "When a mighty blast of violent wind, either coming from without or arising within the earth itself, has suddenly hurled itself into the hollows of the earth, it first roars there tumultuously among the vast caverns, whirling around and around; then its impetuous force, lashed to fury, bursts out and, in so doing, cleaves the earth to its depths and opens a yawning chasm" (578–583), p. 193.

10   Plato, *Volume II: Theaetetus | Sophist*, trans. H. Fowler (New York, 1921), Section 245–249.

11   *Della pittura e della statua di Leonbatista Alberti* (Milan, 1804), p. 28. "La prima cosa nel dipingere una superficie, io vi disegno un quadrangolo di angoli retti grande quanto a me piace, il quale mi serve per un' aperta finestra dalla quale si abbia a veder l'istoria … ."

12   Marsilio Ficino, *The Platonic Theology*, Volume I, trans. M.J.B. Allen, ed. J. Hankins (Cambridge, 2001).

PART TWO
Construction Matters

6

Architectural Encounters between Material and Idea

Paul Emmons

Design is often approached as a process of conceiving an architectural form that is later inserted onto pliant material. Yet architecture is a true co-presence of meaning and material. During design, idea is revealed within, not prior to or in spite of material. Matter is neither only extension nor purely passive: it is a qualitative and cultural condition that describes the nature of being. However it is described, form and material, essence and substance, soul and body, the ultimate condition is their dual presencing. Material need not be conceived as a compromise of form. Architectural drawing aids in this linkage as an intermediary of very subtle material that invites ideation.

ARCHITECTURAL PLATONISM

Architecture has long been dominated by Platonizing assumptions about the relationship between idea and material. It is usually asserted that idea (or the metaphysical) precedes the work (or the physical) in a triumph of idea over matter. It remains a common quip that "architecture is the last stronghold of Platonism," as this approach to the relation between idea and material is remarkably tenacious in a field where design is fundamental.[1] The contemporary architectural language of ideas and forms betrays architecture's platonic sensibilities.[2] Normative architectural practices today perpetuate this idealist approach by beginning with schematic drawings as outlines of shapes in a formal description to which material is later added through design development drawings and specifications. The American Institute of Architects, through its legal documents, especially the B141 as the most commonly used owner–architect agreement in the United States, divides design services into the well-known phases of schematic, design development and construction documents. The first, schematic design, is defined as the building form or outline shown through "conceptual" drawings and a "statistical summary of the design area" without providing material delineations. It is only the later phases of design that call for material specifications, reinforcing the primacy of form over matter.[3]

Perhaps the most important instance and origin of this view of the subservient role of material to idea in architecture is found in the Italian Renaissance discussion of *disegno*, when Neoplatonism was highly influential through such philosophers as the Florentine Marsilio Ficino (1433–1499), who translated the works of Plato and the Neoplatonist Plotinus from Greek into Latin. Ficino defines beauty as the "victory of divine reason over matter" and cautions that forms are corrupted and contaminated when they are embedded in the "bosom" of matter.[4] In his concurrent architectural treatise, Leon Battista Alberti divided the art of building into two aspects: "lineaments and matter, the one the product of thought, the other of Nature; the one requiring the mind and the power of reason, the other dependent upon preparation and selection ...."[5] Alberti's choice of the word lineament reinforced design as first a geometrical line in the mind and only thereafter construction in material by the builder. Giorgio Vasari, founder of the Florentine *Accademia del disegno* in 1563, explains that:

> there is formed in the mind that something which afterwards, when expressed
> by the hands, is called design, we may conclude that design is not other than
> a visible expression and declaration of our inner conception and of that which
> others have imagined and given form to in their idea. ... [Architecture's] designs
> are composed only of lines, which so far as the architect is concerned, are nothing
> else than the beginning and the end of his art, for all the rest ... is merely the
> work of carvers and masons.[6]

The word *disegno*, like its English derivative, means both a physical drawing and a mental concept. Vasari used this dual signification to slide back and forth between them, in order to assert the importance of architecture beyond mere craft. The mental aspects of *disegno* connected it with the liberal arts such as geometry. Ultimately, *disegno* becomes elevated to the primary idea alone (Figure 6.1). Yet, this effort to raise the stature of what are now known as the fine arts by

6.1   Idea.
Giovanni Pietro
Bellori, *Le Vite de'*
Pittori, Scultori et
Architetti Moderni,
Parte prima (Rome:
Mascardi, 1672),
p. 3.

IDEA

emphasizing their intellectual status ultimately had deleterious effects on their long-term relationship to material, which still echoes in architecture today. Because this occurred when the architect had only recently left the construction site and for the scholar's study, the connections with material understanding were still quite present. Over time, the rhetoric was taken as fundamental and the relationship with material and construction became increasingly distant.

The now widely prevalent 3D printer promises to be the ultimate platonic design tool. It makes three-dimensional solid objects of virtually any shape directly from a digital model using an additive process, where successive layers of material are laid down to build up a solid object by printing each new layer on top of the previous one. 3D printing is considered distinct from traditional subtractive techniques, which rely on the removal of material by methods such as carving or drilling. The raw material for most architectural 3D printers is a powder polymer that is almost devoid of any discernible qualitative properties other than extension and offers almost no resistance to manipulation. This technology seems to be very sympathetic with Plotinus's Neoplatonic characterization of things being the "injection of form into matter" where matter is a primal substrate of all materials.[7]

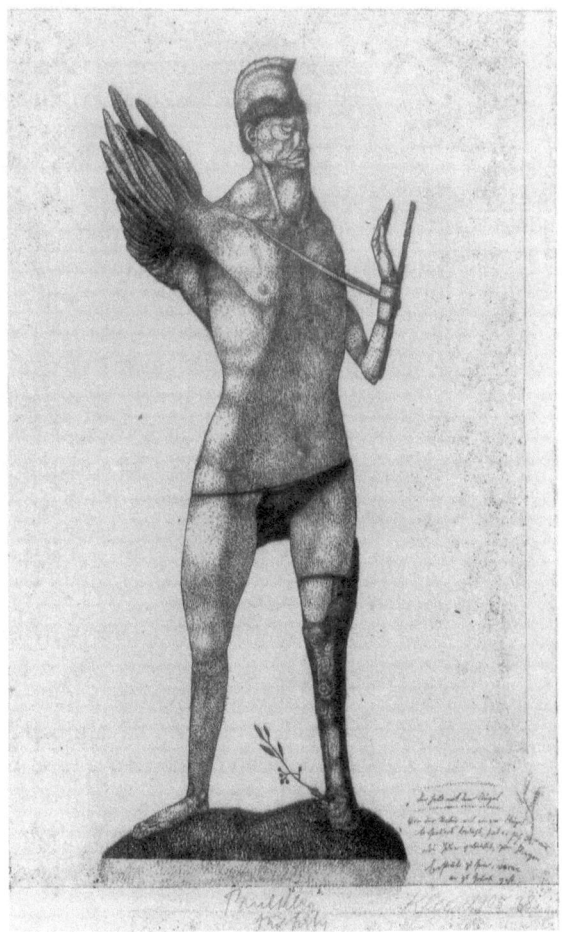

6.2 *The Hero with the Wing*. Paul Klee (1905).

Using 3D printing technology, in architecture professor Neri Oxman's phrasing, "one might turn into physical form any poetry that resides in the mind."[8] In this way, 3D printed products are almost an incarnation of form without material.

IMAGE OF THE ARTIST BETWEEN IDEA AND MATERIAL

Some images of architects and artists explore the relation between idea and material. Three related images are considered here that suggest an alternative view on idea and material. The first is an etching made early in the twentieth century by Swiss-born artist Paul Klee (1879–1940), who later taught the *Vorkurs* (Preliminary or Basic Course) at the Bauhaus. Klee created a series of etchings called *Inventions*, including one titled *Hero with a Wing* (Figure 6.2).[9] Klee's writing on the plate icarously states that, "specially provided by nature with one wing … [he] has conceived the idea that he was destined to fly and has crashed to earth." He is extended into his environment by attaching a wing directly to the shoulder where an arm would be and tying the leg onto a rooted tree stump to emphasize the tension between

6.3   Poverty
Constrained
by Genius.
Paupertatem
summis ingeniis
obesse ne
provehantur.
Andrea Alciato,
*Emblematum
Libellus* (Paris:
Wechel, 1534),
p. 19.

upward and downward movement. Klee's writing about the state of the artist in the twentieth century seems to describe *Hero with a Wing*: "The contrast between man's ideological capacity to move at random through material and metaphysical spaces and his physical limitations, is the origin of all human tragedy. It is this contrast between power and prostration that implies the duality of human existence. *Half winged– half imprisoned, this is man!*"[10] His description speaks to ideas of genius and creativity, since Klee describes artwork as "rendering the invisible visible," which suggests a movement between spirit and material.[11]

A likely source of Klee's image is a 1531 emblem (an enigmatic joining of words and images) by Andrea Alciati called Constrained Genius (Figure 6.3).[12] Alciati's child-like figure or putto in later editions is shown as an adult male. While several other sources have been proposed for inspiring Klee's etching, there are a number of aspects of this drawing that link it with this emblem.[13] Besides the specific combination of image and text (both a motto and an epigram), the image itself, like an emblem, consists of a distinctive figure with memorable attributes on a small base. Klee, whose education in Bern was in the classics, compared his *Inventions* to "the old woodcuts" of the Renaissance that have "allegorical subject matter."[14] Klee's image reinterprets the wing and rock attributes as conditions of the figure itself in place of its raised arm and lowered leg. In Alciati's emblem, the motto is "poverty hinders the greatest talents from advancing."[15] The notion of poverty in Alciati's emblem goes beyond current conceptions of insufficient monetary income and instead describes the human condition as one that is inevitably incomplete and partial. Ivan Illich describes "modernized poverty" as that which turned "lack" into a "need" that must be addressed by economic consumption.[16] Klee's description of the artist as "hero" caught between the physical and the metaphysical is very similar to Alciati's conception and reflects similar thinking whether or not Klee's *Hero* directly derives from it. The emblem of constrained genius, particularly popular throughout the Renaissance, appeared in many guises.[17]

THE FRONTISPIECE OF WALTHER HERMANN RYFF

The constrained genius emblem from Alciati was certainly a direct antecedent for a Renaissance architectural frontispiece in works by Walther Ryff (or Rivius, c.1500–1548), who had a brief but prolific career as an author and translator of popular scientific works.[18] Ryff's frontispiece for his amalgam on architecture, *Der Furnembsten* or *Architectur* (1547/1548), also appears in his *Vitruvius Teutsch* (1548), the first German translation of Vitruvius.[19] Since most of Ryff's works are adapted from authors writing in Italian and Latin, Ryff earned the reputation of "archplagiarist."[20] While today it would be more accurate to describe him as translator and compiler, ideas of literary property rights were not so clearly defined and recent scholars have defended Ryff's pillaging of other authors, whom he did at least identify, as fairly typical of his time.[21] Most importantly, Ryff did not simply copy his sources; he added, deleted and altered works in new ways to adapt foreign texts to his German audience, calling his work "Germanization."[22] Like the texts, Ryff's images are also almost entirely adapted from other sources to reflect the German milieu. An image of ancient Halicarnassus is revised to look like a northern European city and Ryff's version of the ideal Vitruvian city adds a northern windmill that makes the key idea of city-forming wind concretely visible. Ryff's frontispiece is not merely a copy of some other work; it is an intentional assemblage of elements to create a statement about architecture.[23]

The putto in Ryff's frontispiece is the spirit of architecture (Figure 6.4). As pneumatic creatures, putti are usually winged because they have a highly rarified material presence, as Ryff's putto with curly hair and billowing clothing vividly expresses.[24] Ryff similarly describes architects' "subtle skillfulness" to "put forth and pre-figure" designs as "ingenious inventions."[25] Like many representations of Lady Architecture, the putto's sleeves are raised above the elbows to show readiness for work, while his eyes are gazing upward.

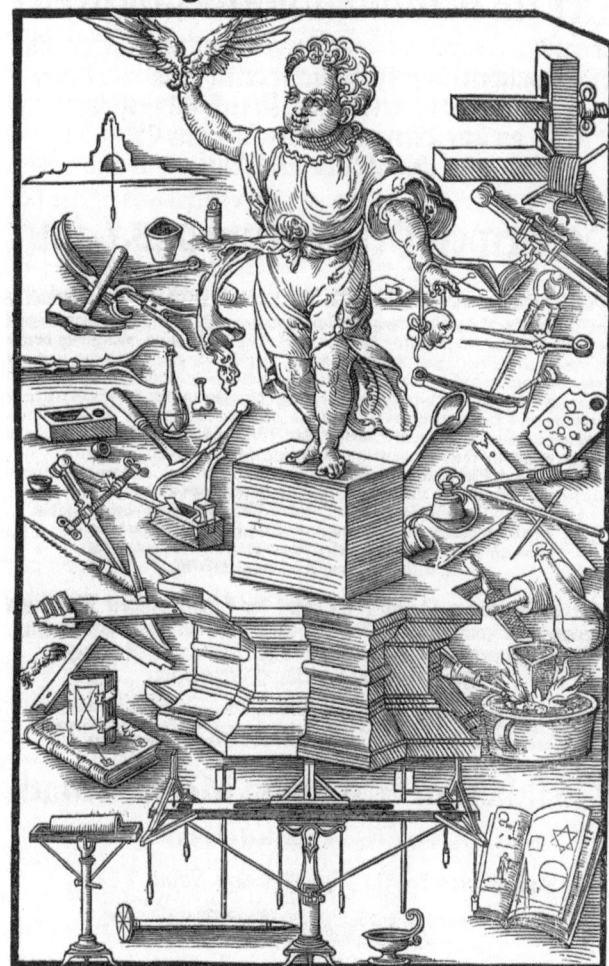

Viuitur ingenio, cætera mortis erunt.

Aurum probatur igni, ingenium uero Mathematicis.

6.4   Frontispiece, Gualtherus Hermenius Rivius, *Der furnembsten notwendigsten der gantzen Architectur* (Nürnberg: Petreius, 1547), verso of title page.

In the original Alciati emblem, the string is tied around the wrist but looped around the rock, and in later versions it is irremovably chained around the wrist. In Ryff, however, the rock is tied with a knotted string, but only looped around the wrist so it could be easily removed by the wearer. Similarly, the wings are grasped by the hand of the putto, not integral with its body. All this suggests that the so-called poverty of material is a knowing choice for the spirit of architecture.

The emblem of constrained genius affirms architecture as a *scientia media*, or in-between science in using geometry as a realm of the mind while exercising it practically with physical materials upon the earth. In this way, architecture is mediate between the metaphysical and the physical. Ryff distinguishes his approach from the abstract approach of the mathematician, rather focusing on those who work with material objects: "The mathematician involves himself in fashioning groups and forms of things only mentally; he ignores their material reality: we shall present the subjects we treat as perceptible."[26] A common opposition of figures in architectural treatise frontispieces is the maidens Theory and Practice.[27] Youthful Theory, dressed in blue, gazes aloft and holds her compasses pointing upwards, while mature Practice, dressed in brown, points her compasses down to the earth. Ryff uniquely joins them into a single figure. The two diagonal arms are united by the vertical central axis of the body with the center of vision near the navel or origin. The central line of the figure is also the center of the image and it aligns with the center of the water level at the bottom of the page, suggesting balance and harmony in graphic layout as well as in the iconography of the elements. Rather than starting with an idea or form and later imposing it on material, the idea emerges within the understanding of material. Through the frontispiece, one can grasp Ryff's project as a subtle unification of craft practice and Vitruvian theory. All of these reinforce the resolution of opposites through the architect's applied art.

The base upon which the putto stands is original to the frontispiece.[28] No other Constrained Genius stands upon a cube. Another of Alciati's emblems, "Art helping Nature," shows Fortune balancing on a sphere while Mercury, as god of the arts, sits on a cube, suggesting that art can correct the uncertainty of nature.[29] Generally, the cube is used in emblems to describe stability, permanence and firmness.[30] Unlike Constrained Genius, representations that consistently show one foot down and one foot raised, Ryff's figure has both feet upon the cube to show the importance and fixity of knowledge for the spirit of architecture.

The cube rests upon an eight-pointed plinth, made of a cube rotated within another cube. Classical molding distinguishes between the two intertwined forms. It is probably related to demonstrations of geometrical skill in Albrecht Dürer's *Melencolia I* (1514) and in the emerging tradition of German perspective manuals.[31] Since the platonic solids represent material elements, their internal rotation demonstrated transmutation.[32] In Caparoli's 1536 Vitruvian treatise, intersecting tetrahedrons and octahedrons as fire and air show the process of steam generation (Figure 6.5).[33] Ryff's precisely finished base signals the effectiveness of the architect's tools and contrasts with the rough stone in the putto's hand as between nature and art as *poeises* or knowledgeable making. Of the three stone elements, the rough rock in the hand is highest, then the pure cube and finally the architectural plinth. This suggests the transformation of rough rock through geometry to constructed architecture through geometry that moves from above to below. This order inverts the Neoplatonic order of dematerializing while moving upward. This inversion suggests a different relation between idea and material in the frontispiece. The importance of knowledge and idea in the direct engagement of material asserts that material is not a mere receptacle of form, but an active aspect of design.

A book about a trade or a profession customarily shows its tools on the title page.[34] More than a catalog, however, the tools express the expertise latent within them.[35] In the introduction to *Vitruvius Teutsch*, Ryff emphasized the importance of the development of tools to bring forth the excellent art of architecture and even that they are the basis of all the other arts.[36] The many tools and instruments scattered across Ryff's frontispiece seem to be as if in another plane from the rest,

LA FIGVRA DE LE AEOLIPILE CIOE PALLE.

6.5   La Figure de le Aeolipile cioe palle. Gianbatista Caporali, *Marco Vitruvio Pollione, De Architecturea Libri I–V* (Perugia: Bigazzini, 1536), p. 32r.

so that unlike the perspective image of the figure and base, the tools appear more to remain on the surface of the paper. By not receding into the distance, the tools are immediately graspable by the reader, as equipment available present-at-hand that are not mere things, but already suggesting certain kinds of work. The frontispiece's tools are almost all related to geometry, both theoretical and practical, spanning from drawing instruments to tools for translating drawings into buildings. The cube, compass, inkwells and writing materials are more related to the pure geometry of theory. The levels, surveying equipment, molding plane, set square and hammer realize geometry in building and craft practices. Some tools, like compasses and plumb lines, often apply to both. The bellows, crucible, tongs and many of the bottles are related to alchemy. The title page of *Vitruvius Teutsch* states that the book is for "all artistic handworkers, builders, wood workers, painters, sculptors, goldsmiths, carpenters and others who use artistic titles and tools."[37] The bellows in Ryff's frontispiece protrudes from behind the plinth and points toward a flame around a crucible. The bellows, also important to alchemy, is literally and metaphorically tied to air and spirit. The bellows is even considered animate, a miraculous object that has a force of its own, beyond the operator.[38] In this way, the bellows is a machine of the spirit. The crucible is used for heating metals and is shown with a fire like an alchemical experiment. This may be a reference to the motto included at the foot of the frontispiece: *Aurum probatur igni, ingenium uero Mathematicis* (Gold is assessed with fire, genius with mathematics).[39] Ryff describes architecture in the introduction to *Vitruvius Teutsch* as an "ingenious mathematical art."[40] The proverb also has a strong alchemical sense that is reinforced in Ryff's frontispiece by its proximity to the crucible for melting metals. The motto emphasizes the important relation of genius, another name for a putto, to both material and idea.

There are several books among the instruments showing the importance of learning to the practice of architecture. The two books on the raised arm side are closed and clasped, suggesting hidden and divine knowledge; while on the other lowered arm side the two open books are the discursive human knowledge of the arts.[41] One open book has images of applied geometry dominated by a pair of overlapping equilateral triangles. The six-pointed star made of two equilateral triangles is known in alchemy as the Seal of Solomon. It represents the joining together of the elements "so that all is unified in perfect balance."[42] From the opposition of fire (upward triangle) and water (downward triangle) the active/passive and *forma/materia* becomes a union of opposites. This union of heaven and earth, spirit and flesh is conceived as transporting heaven to earth. For the putto, as suggested in Alciati's text, the movement upwards with wings is alchemically related to the levity of fiery air and the downward movement of the stone to the gravity of earth. His navel, at the midpoint and emphasized by the uniting knot in his cumberbund, is reaching up to the ideal and down to material. Similarly, the crucible refers to a Christological path of purification. Alchemically, the crucible was used to make a base metal molten on which the elixir was projected for the final transmutation to take place.[43] Many of the frontispiece's elements identified above indicate the mixed and mediating nature of architecture. The geometrical rotation of the cube to create the plinth is also present in the Seal of Solomon figure.

This frontispiece describes a largely Aristotelian philosophical approach to the creative relation between idea and material in design. The reevaluation of Aristotle in the Renaissance led to daring departures within the Aristotelian framework, including early modern science.[44] Pietro Pomponazzi of Mantua, who was teaching at the University in Padua known for its progressive Aristotelianism, wrote a commentary on *de Anima* (the soul) that humans are a mean between material and intellect. Neither materialist nor idealist, Pomponazzi argued that the intellect is the mean between eternal and non-eternal as the first of material forms because "it is in this flesh that we can behold truth." The intellect can only act within the body and its corporeal sense images. He attacks the Platonic view by arguing, "if soul and body have no more unity than oxen and a cart, there would be two men joined together in me."[45] Ryff's treatise, which repeatedly cites Aristotle, includes images showing geometric diagrams imposed on perspectival scenes to explain physical movement. It was this achievement that is typically credited to Galileo a century later. The idea or form is not prior but occurs within the thing.[46] The materialization of idea is also evident in Ryff's repeated use of the phrase "sense-full reason" as good knowledge. He concludes this discussion by noting that "this splendid art is nothing else than an imitation or artificial reproduction of the divine acting through which heaven and earth including all so therein can be grasped, formed and created and richly decorated."[47]

The spirit understood as in-between idea and material is readily compared to the architectural drawing as a body of subtle material that embodies an idea. Ryff describes drawing as "subtle skillfulness" and refers to the architect's task as to "put forth and pre-figure," and adds that like Aristotle on signs, he who has made such a thing of the hand can show it. When the angelic, subtle body of the drawing is understood to be contiguous with the physical body of the building, there is a chiasmatic relation between them, the real presence of both something and its negation, simultaneously, the making palpable of the non-sensible. This is only achieved when working through both material and idea, not forcing one onto the other.

Ryff used Vitruvian doctrine to argue for architecture in the north to be a liberal art along with other mathematical arts and combined this with his practical orientation.[48] He furthered the *architector doctus* as a cultured, educated architect that also required extensive technical and practical skills. It was not unusual to draw comparisons between the divine creator and architectural constructions. But Ryff wrote of heavenly creation as if describing a manipulator of material that "each piece or part, together with its action, was planned in advance through the marvelous wisdom and forethought of God, who is the most superior *master craftsman* of nature."[49] Rather than God as in Genesis creating the world from nothing, Ryff presents divine creation as demiurgic, a manipulator of matter.

SACRAMENTAL SPIRIT AND MATERIAL

In Ryff's lifetime, enormous religious debates occurred during the Reformation in his German homeland. Although Martin Luther (1483–1546) was not against all religious images, iconoclasm forced many artists to find new sources of patronage.[50] The artists who illustrated Ryff's architectural works also created anti-Catholic images.[51] Part of the motive for Ryff's publications may have been to advance a more secularized classical imagery rather than the still-dominant gothic in northern European religious buildings.[52] The most profound encounter between spirit and matter reflected in these debates occurs with theophagy, the Sacrament of Holy Communion.[53] To explain the relation between the immaterial and material in how bread and wine become Christ's body and blood, Scholastic theologians developed the theory of *transubstantiation*, that the substance of the bread is transformed into the body or flesh of Christ with only the accidental qualities of bread, such as form and color, remaining after its consecration by a priest. The real presence of the divine is in this way contained in the external appearance of the bread. This occurs not as mere change but as the conversion of one thing into another.

For Luther, the sacrament is a "visible word" because it combines the Gospel with material (bread and wine). He affirmed the real presence of Christ in the bread but proposed an alternative explanation with *consubstantiation*. The substance bread and the substance of the body of Christ co-exist together without one becoming the other. Luther explained that Christ's body is consubstantially with or under the unconsecrated bread and that the Sacrament "fastens" (*fasst*) spirit to matter. The German word *fassen* connotes both physical and mental "grasping."[54] A similar third alternative in the Anglican Church called *impanation* was also proposed to explain the real presence of Christ in consecrated bread. Impanation holds that Christ becomes bread through a hypostatic union, like incarnation where god becomes human. With impanation, Christ is incarnated into the bread so that his human body is what becomes present in the bread. Christ is in the bread without the change of the substance of the bread into his substance. All three of these explanations accepted a real divine presence in the material. A radically different alternative was put forth by Ulrich Zwingli (1484–1531) and the Enthusiasts, that the sacrament is merely a symbol of union without any real divine presence. This extreme position was rejected not only by the Catholic Church but also by Luther and other protestant sects.

While architecture and theophagy may appear at first entirely unrelated, many of the philosophical issues about idea and material are very similar. Perhaps it is for this reason that one finds examples of people with those overlapping interests in both time periods under consideration here. In the Renaissance, Daniele Barbaro, patron of Palladio and author of a definitive translation and commentary on Vitruvius, discussed the issue of transubstantiation at the Council of Trent.[55] Early in the twentieth century about when Klee made his etching, American architect and neo-gothic apologist Ralph Adams Cram lectured and published on transubstantiation.[56] Philosopher Hans-Georg Gadamer used the reformation controversy over the real presence of Christ in consecrated bread and wine to illustrate how works of art are meaningful within their material presence as neither materialism nor idealism.[57]

To explore a range of real encounters between material and idea in architecture, consider the following brief examples as three sorts of approaches to an architectural theophagy. The idea of a mobile holy place marked by a baldachin or canopy of tasseled fabric on four wooden staves was transubstantiated into gilded bronze by Gian Lorenzo Bernini for St. Peters in Rome. Bernini's *Baldacchino* (1624–1633) was made from over one hundred thousand pounds of bronze (largely taken from the Pantheon's ancient porch beams) and cast by the same foundry and process that made canons for the Vatican army. The interiors of the columns were filled with mortar to support the structure.[58] Despite its enormous weight and height of almost 29 meters tall, it creates the feeling of a lightweight structure. Gilded bronze, two of the most precious metals, was often used in reliquaries holding sacred remains. The accidental qualities of the fabric and wood structure remain while it is transubstantiated into bronze and gold.

The idea of artistic decoration deriving from construction was consubstantiated by Otto Wagner in his Church of St. Leopold am Steinhof (Vienna, 1904–1907). Like his Post Office Savings Bank, these projects were designed and built about the time that Klee made the *Hero with Wing* etching. Both buildings have marble panels with carefully designed bolt caps that let the stone appear like the thin veneer that it is and the spacing of the caps creates an ornamental pattern. At one moment in the design, Wagner proposed gilding the caps. The stone is set into a mortar bed on the brick supporting wall so the metal ties were only actually structural during the time that the mortar was hardening.[59] Like consubstantiation, the two substances co-exist with each being present and are fastened together to form a bond.

At almost the same time, Frank Lloyd Wright impanated space in Unity Temple (1905) in Oak Park, Illinois. Wright imagines wood as sawn, dimensional lumber. In describing the "orderly piles of freshly cut and dried timber disappearing into the mills to be gored and ground and torn and hacked into millwork," Wright argues that dimensional lumber is a preferred primal condition of wood, calling it a "stick" at once natural like timber and subject to art like wood.[60] Like impanation, wood slats both freestanding and moving across plaster surfaces articulate the flowing spatial idea in the Temple in a hypostatic union of idea and matter, space and material. These three cases suggest the different sorts of relationships between idea and material that can be manifest, but they all emphasize a real presence, none of them accepting merely symbolic relations. Architects generally all reject the use of one material to imitate another, without its real presence being made manifest.[61] Each of the three examples highlights the mystery of material imagination in architectural idea-making. However achieved, the deeper significance of material is accomplished with the union of idea and material in architecture where the architect is the figure of the interval.

CONCLUSION: TOWARD REAL PRESENCE IN ARCHITECTURE

Today, matter is not the same as it was in pre-modern times. Now, spirit is divorced from lifeless matter. Earlier, spirit was present within matter as its necessary complement. For Ficino, matter is animate, pregnant with layers of significance. In between the soul and the body is the spirit that provides an "ineffable union" between soul and body. Similarly, Luther in 1521 conceived of the Church as neither pure spirit nor pure matter, but a hybrid, a *corpus mixtum*, where both modalities obtained and "where the beyond attaches to the here and now."[62] The architectural drawing, itself a spiritual existence of very subtle matter and very heavy idea, provides the bridge that actively unites idea and material during design. Without this essential relation, the rhetoric of the Italian Renaissance *disegno* becomes quickly misunderstood as isolation and prioritization of idea. When drawing is conceived as merely recording already determined ideas, the potential of this essential design tool is lost. Like Plato's characterization of the *Statesman* as mediating between the ideal Good and shifting circumstances on earth, so the architect intertwines the theoretical and the material.[63]

Because of the dominance of the formal imagination in architecture, the materic nature of building is often unduly limited within the architect's imagination. The architect as Constrained Genius is not burdened with shackles to be lamented; it is the condition of human existence—embodiment—that defines the delightfully complex art of architecture. The architect's material imagination simultaneously invites rising in oneiric reverie and feeling the gravity of physical weight. As Marco Frascari has suggested, "Material imagination is both at the same time imagination of the material and the material of imagination, in other words, the solid body and the subtle body simultaneously."[64] Ryff's frontispiece can continue to inspire critical and creative thought and practices with the architectural encounter between idea and material.

Ultimately, the question is one of *presence*. Materiality becomes a real presence when the "spirit" (genius) of architecture is perceived as being co-terminal between idea and matter. Presence is not a fracture between appearance and essence, sensible and intelligible; it is an original experience that is always already caught in a fold of meaningful material. Real presence is a direct relation with facture or making that opens a world, as Paul Klee suggested, making visible the invisible.

Klee's etching describes Constrained Genius as a tragic character, due to the apparent hopelessness of uniting idea or the "cosmic" and body or the "earthbound" in the modern world. Klee's figure apparently no longer offers any hope of wholeness or resolution of the divergent forces pulling it apart. Klee emphasizes the dichotomy of human existence: "Man is half a prisoner, half borne on wings. Each of the two halves perceives the tragedy of its halfness by awareness of its counterpart."[65] Yet, even this "tragedy of spirituality" is tragicomic, and its humor points to the tragicomic hero with the wing as a contemporary Don Quixote. Perhaps a way forward is exactly through its tragicomic nature; with the ability to laugh at ourselves we can take a critical stance to our modern technology-driven ideology.

BIBLIOGRAPHY

Alberti, Leon Battista, *On the Art of Building in Ten Books*, trans. J. Rykwert, N. Leach and R. Tavernor (Cambridge, MA: MIT Press, 1988).

Alciati, Andrea, *A Book of Emblems; The* Emblematum Liber *in Latin and English*, trans. J. Moffitt (Jefferson, NC: McFarland, 2004).

Benzing, Josef, "Walter H. Ryff und sein literarisches Werk," *Philobiblon* 2 (1958).

Bowen, Barbara, "Mercury at the Crossroads in Renaissance Emblems," *Journal of Warburg and Courtauld Institutes* 48 (1985), pp. 222–9.

Burckhardt, Titus, *Alchemy: Science of the Cosmos, Science of the Soul*, trans. W. Stoddart (Louisville, KY: Fons Vitae, 1997 [1960]).

Caporali, Gianbatista, *Marco Vitruvio Pollione, De Architectura, Libri I–V* (Perugia: Iano Bigazzini, 1536).

Corbett, Margery, "The Architectural Title-page; an Attempt to Trace its Development from its Humanist Origins up to the Sixteenth and Seventeenth Centuries, the Heyday of the Complex Engraved Title-page," *Motif* 12 (1964), pp. 48–62.

Cram, Ralph Adams, *Gold, Frankincense and Myrrh* (Boston, MA: Marshall Jones, 1919).

Crowther, Kathleen, *Adam and Eve in the Protestant Reformation* (Cambridge: Cambridge University Press, 2010).

Demkin, Joseph (ed.), *The Architect's Handbook of Professional Practice, Student Edition* (New York, NY: AIA and John Wiley & Sons, 2002).

Dempsey, Charles, *Inventing the Renaissance Putto* (Chapel Hill, NC: University of North Carolina Press 2001).

Eamon, William, *Science and the Secrets of Nature: Books of Secrets in Medieval and Early Modern Culture* (Princeton, NJ: Princeton University Press, 1994).

Eliade, Mircea, *The Forge and the Crucible: The Origins and Structures of Alchemy* (Chicago, IL: University of Chicago Press, 1978).

Ficino, Marsilio, *Platonic Theology*, trans. Michael Allen and John Warden, 6 vols. (Cambridge, MA: Harvard University Press, 2001–2006).

Forty, Adrian, *Words and Buildings: A Vocabulary of Modern Architecture* (New York: Thames & Hudson, 2004).

Fowler, A.D.S., "Emblems of Temperance in the Faerie Queene Book II," *Review of English Studies* 11.42 (May, 1960), pp. 143–9.

Franciscono, Marcel, "Paul Klee's Italian Journey and the Classical Tradition," *Pantheon* 32.1 (January/March, 1974), pp. 54–64.

Frascari, Marco, "Maidens 'Theory' and 'Practice' at the Sides of Lady Architecture," *Assemblage* 7 (1988), pp. 15–27.

Gadamer, Hans-Georg, *The Relevance of the Beautiful*, trans. N. Walker (Cambridge: Cambridge University Press, 1986 [1977]).

Harig, Gerhard, "Walter Hermann Ryff und Nicolo Tartaglia: Ein Beitrag zur Entwicklung der Dynamik im 16. Jahrhundert," in *Physik und Renaissance; zwei Arbeiten zum Entstehen der klassischen Naturwissenschaften in Europa* (Leipzig: Geest & Portig, 1984), pp. 13–36.

Heuer, Christopher, *The City Rehearsed: Object, Architecture and Print in the Worlds of Hans Bredeman de Vries* (London: Routledge, 2009).

Illich, Ivan, *The Right to Useful Unemployment* (New York, NY: Marion Boyars, 1978).

Ionescu, Cristina, "Dialectical Method and Myth in Plato's Statesman," *Ancient Philosophy* 34 (2014), pp. 1–18.

Jachmann, Julian, *Die Architekturbücher des Walter Hermann Ryff: Vitruvrezeption im Kontext mathematischer Wissenschaften* (Stuttgart: Ibidem, 2006).

Jamnitzer, Wenzel, *Perspectiva Corporum Regularium* (Nuremberg, 1568).

Johnson, Paul-Alan, *Theory of Architecture: Concepts, Themes, Practices* (New York, NY: Van Nostrand Reinhold, 1994).

Kirwin, W. Chandler, *Powers Matchless: The Pontificate of Urban VIII, the Baldachin, and Gian Lorenzo Bernini* (New York, NY: Peter Lang, 1997).

Klee, Paul, *Pedagogical Sketchbook*, trans. S. Moholy-Nagy (London: Faber and Faber, [1925] 1953).

Klee, Paul, *Notebooks, Volume 1: The Thinking Eye*, trans. Ralph Manheim (London: Percy Lund, 1961).

Klibansky, Raymond, Erwin Panofsky and Fritz Saxl, *Saturn and Melancholy: Studies in the History of Natural Philosophy, Religion and Art* (New York, NY: Basic Books, 1964).

Koerner, Joseph Leo, *The Reformation of the Image* (Chicago, IL: University of Chicago Press, 2004).

Laven, Peter, *Daniele Barbaro, Patriarch Elect of Aquileia; with Special Reference to His Circle of Scholars and to His Literary Achievement* (University of London: Ph.D. dissertation, 1957).

Loos, Adolf "The Principle of Cladding," in Michael Mitchell (trans.), *On Architecture* (Riverside, CA: Ariadne Press, 1995), pp. 42–7.

Mallgrave, Harry Francis, "Introduction," in Otto Wagner, *Modern Architecture: A Guidebook for His Students to This Field of Art* (Santa Monica, CA: Getty Center for the History of Art and Humanities, 1988), pp. 1–51.

Mallgrave, Harry Francis, "Introduction," in Harry Francis Mallgrave, Gerald Beasley, Claire Baines, et al. (eds), *The Mark J. Millard Architectural Collection, Volume III, Northern European Books, Sixteenth to Early Nineteenth Centuries* (Washington, DC: National Gallery of Art, 2002), pp. 1–61.

Mallgrave, Harry Francis, Gerald Beasley, Claire Baines, et al., *The Mark J. Millard Architectural Collection, Volume III, Northern European Books, Sixteenth to Early Nineteenth Centuries* (Washington, DC: National Gallery of Art, 1998).

Maué, Hermann, "Nuremberg's Cityscape and Architecture," in *Gothic and Renaissance Art in Nuremberg, 1300–1550* (Munich: Presetel, 1986), pp. 27–50.

Miedema, Hessel, "The Term Emblema in Alciati," *Journal of the Warburg and Courtauld Institutes* 31 (1968), pp. 234–50.

Oechslin, Werner, "Vitruvianismus in Deutschland," in *Architekt und Ingenieur: Baumeister in Krieg und Frieden* (Wolfenbüttel: Herzog August, 1984), pp. 53–76.

Oxman, Neri. Available at: http://singularityhub.com/2012/06/04/3d-printing-is-the-future-of-manufacturing-and-neri-oxman-shows-how-beautiful-it-can-be/ [accessed March 13, 2014].

Oxman, Neri, "Structuring Materiality: Design Fabrication of Heterogeneous Materials," in *The New Structuralism: Design, Engineering and Architectural Technologies, Architectural Design*, 80.4 (New York, NY: Wiley, July/August, 2010), pp. 78–85.

Pacioli, Luca, *De divina proportione, Facsimile of the ms. in the Bibliteca Ambrosiana di Milano (1458)* (Milan: Silvana, 1982).

Panofsky, Erwin, *Idea: A Concept in Art Theory*, trans. J. Peake (Columbia, SC: University of South Carolina Press, 1968).

Pomponazzi, Pietro, "On the Immortality of the Soul," in *The Renaissance Philosophy of Man*, edited by Ernst Cassirer, et al. (Chicago, IL: University of Chicago Press, 1948), pp. 280–384.

Randall, John Herman Jr., *The School of Padua and the Emergence of Modern Science* (Padua: Editrice Antenore, 1961).

Rivius, Gualtherus Hermenius, *Der furnembsten notwendigsten der gantzen Architectur angehorigen mathematischen und mechanischen Kunst eygentlicher Bericht und verstendliche Unterrichtung* (Hildesheim: George Olms, 1981 [1547]).

Rivius, Gualtherus Hermenius, *Vitruvius Teutsch* (Nuremberg: Johann Petreius, 1548).

Roberts, Gareth, *The Mirror of Alchemy: Alchemical Ideas and Images in Manuscripts and Books from Antiquity to the Seventeenth Century* (Toronto, ON: University of Toronto Press, 1994).

Rosenthal, Mark, "The Myth of Flight in the Art of Paul Klee," *Arts Magazine* 55.1 (September, 1980), pp. 90–94.

Röttinger, Heinrich, *Die Holzschnitte zur Architektur und zum Vitruvius Teutsch des Walther Rivius* (Strassburg: Heitz, 1914).

Sambucus, Johannes, *Emblemata cum aliquot Nummi antiqui operis* (Antwerp: Christopher Plantin, 1564).

Smith, Jeffrey, *Nuremberg: A Renaissance City, 1500–1618* (Austin, TX: University of Texas Press, 1983).

Smith, Preserved, *A Short History of Christian Theophagy* (Chicago, IL: Open Court, 1922).

Stoer, Lorenz, *Geometria et Perspectiva* (Nuremberg, 1567).

Tawa, Michael, "Unity and the Idea," *Transition* 9 (March, 1982), pp. 32–7.

Vasari, Giorgio, *Vasari on Technique; Being the Introduction to the Three Arts of Design, Architecture, Sculpture and Painting, prefixed to the* Lives of the Most Excellent Painters, Sculptors and Architects, trans. L. Maclehose (New York, NY: Dover, 1960).

Vesalius, Andreas, *De humani corporis fabrica libri septem* (Basel, 1543).

Wade, David, *Fantastic Geometry: Polyhedra and the Artistic Imagination in the Renaissance* (London: Squeeze, 2012).

Watson, Stephen, "Gadamer, Aesthetic Modernism, and the Rehabilitation of Allegory: The Relevance of Paul Klee," *Research in Phenomenology* 34 (2004), pp. 45–72.

Wood, Christopher, "The Perspective Treatise in Ruins: Lorenz Stoer, *Geometria et perspectiva*, 1567," in Lyle Massey (ed.), *The Treatise on Perspective, Published and Unpublished* (Washington, DC: National Gallery of Art), pp. 235–58.

Wright, Frank Lloyd, "In the Cause of Architecture: The Meaning of Materials – Wood (May, 1928)," in Frederick Gutheim (ed.), *In the Cause of Architecture, Frank Lloyd Wright* (New York, NY: Architectural Record Books, 1975), pp. 179–88.

NOTES

1   Paul-Alan Johnson, *Theory of Architecture: Concepts, Themes, Practices* (New York, NY, 1994), p. 244; quoted in Adrian Forty, *Words and Buildings: A Vocabulary of Modern Architecture* (New York, NY, 2004), p. 136.

2   Michael Tawa, "Unity and the Idea," *Transition* 9 (March, 1982), pp. 32–7.

3   Joseph Demkin (ed.), *The Architect's Handbook of Professional Practice, Student Edition* (New York, NY, 2002), p. 435.

4   Marsilio Ficino, *Platonic Theology*, trans. Michael Allen and John Warden (Cambridge, MA, 2001), I.III.15; quoted in Erwin Panofsky, *Idea: A Concept in Art Theory*, trans. J. Peake (Columbia, SC, 1968), p. 53.

5   Leon Battista Alberti, *On the Art of Building in Ten Books*, trans. J. Rykwert, N. Leach and R. Tavernor (Cambridge, MA, 1988), Prologue, p. 5.

6   Giorgio Vasari, *Vasari on Technique; Being the Introduction to the Three Arts of Design, Architecture, Sculpture and Painting, prefixed to the* Lives of the Most Excellent Painters, Sculptors and Architects, trans. L. Maclehose (New York, NY, 1960), pp. 205–7.

7   Plotinus, *Enneads*, 5. 8. 1; quoted in Panofsky, *Idea*, 27.

8   Neri Oxman. Available at: http://singularityhub.com/2012/06/04/3d-printing-is-the-future-of-manufacturing-and-neri-oxman-shows-how-beautiful-it-can-be/ [accessed March 13, 2014]. Neri Oxman, "Structuring Materiality: Design Fabrication of Heterogeneous Materials," in *The New Structuralism: Design, Engineering and Architectural Technologies, Architectural Design*, 80.4 (New York, NY, July/August 2010), pp. 78–85.

9   Paul Klee, *Der Held mit dem Flügel (The Hero with the Wing)*, etching and drypoint, 1905.

10   Paul Klee, *Pedagogical Sketchbook*, trans. by Sibyl Moholy-Nagy (London, [1925] 1953), pp. 37, 54.

11   Stephen Watson, "Gadamer, Aesthetic Modernism, and the Rehabilitation of Allegory: The Relevance of Paul Klee," *Research in Phenomenology* 34 (2004), pp. 45–72, 48.

12   Andrea Alciati, *A Book of Emblems; The* Emblematum Liber *in Latin and English*, trans. J. Moffitt (Jefferson, NC, 2004), Emblem 120, p. 143.

13   Various proposed sources by art historians for Klee's image: Goya's *Proverbios*, Cervantes' *Don Quixote*, Icarus of Greek myth and fragmented ancient sculpture such as the Doryphorus.

14   Marcel Franciscono, "Paul Klee's Italian Journey and the Classical Tradition," *Pantheon* 32.1 (January/March 1974), pp. 54–64, 55–6. Mark Rosenthal, "The Myth of Flight in the Art of Paul Klee," *Arts Magazine* 55.1 (September, 1980), pp. 90–94.

15   *Paupertatem summis ingeniis obesse, ne provehantur*. The complete translated text is: "My right hand holds a stone, my other hand bears wings. As the feathers lift me, so the heavy weight drags me down. With my intellect I could be soaring among the highest peaks, if envious poverty did not pull me down." Alciati, p. 143.

16   Ivan Illich, *The Right to Useful Unemployment* (New York, NY, 1978), pp. 8–11.

17   "*Physicae ac Metaphysicae differentia*," in Johannes Sambucus, *Emblemata cum aliquot Nummi antiqui operis* (Antwerp, 1564), p. 74.

18   Josef Benzing, "Walter H. Ryff und sein literarisches Werk," *Philobiblon* 2 (1958). Julian Jachmann, *Die Architekturbücher des Walter Hermann Ryff: Vitruvrezeption im Kontext mathematischer Wissenschaften* (Stuttgart, 2006).

19   Gualtherus Hermenius Rivius, *Der furnembsten notwendigsten der gantzen Architectur angehorigen mathematischen und mechanischen Kunst eygentlicher Bericht und verstendliche Unterrichtung* (Hildesheim, 1981 [1547]), Title Page verso; Book I, LIXv; Book III, IVr. The second edition was published as *Der Architectur furnembsten, notwendigsten, angehörigen Mathematischen und Mechanischen künst* (Nuremberg, 1548). Ibid., *Vitruvius Teutsch* (Nuremberg, 1548), XI. Harry Francis Mallgrave, Gerald Beasley, Claire Baines, et al., eds., *The Mark J. Millard Architectural Collection: Northern European Books, Sixteenth to Early Nineteenth Centuries, Volume III* (Washington, DC, 1998), pp. 329–30.

20   Andreas Vesalius, *De humani corporis fabrica libri septem* (Basel, 1543), Book III, preface.

21   William Eamon, *Science and the Secrets of Nature: Books of Secrets in Medieval and Early Modern Culture* (Princeton, NJ, 1994).

22   Rivius, *Vitruvius Teutsch*, Introduction [n.p.]. Translated by Ulrike Altenmüller.

23   Jeffrey Smith, *Nuremberg: A Renaissance City, 1500–1618* (Austin, TX, 1983), p. 233. Heinrich Röttinger, *Die Holzschnitte zur Architektur und zum Vitruvius Teutsch des Walther Rivius* (Strassburg, 1914), pp. 41–51.

24   Charles Dempsey, *Inventing the Renaissance Putto* (Chapel Hill, NC, 2001), pp. 43–5.

25   Rivius, *Vitruvius Teutsch*, Introduction [n.p.]. Translated by Ulrike Altenmüller.

26   Ryff, trans. in Christopher Heuer, *The City Rehearsed: Object, Architecture and Print in the Worlds of Hans Bredeman de Vries* (London, 2009), p. 187, n. 70.

27   Marco Frascari, "Maidens 'Theory' and 'Practice' at the Sides of Lady Architecture," *Assemblage* 7 (1988), pp. 15–27.

28   There is "no model known" for the "cube and gothic base." Jachmann, p. 114.

29   Alciati, emblem 98. Barbara Bowen, "Mercury at the Crossroads in Renaissance Emblems," *Journal of Warburg and Courtauld Institutes* 48 (1985), pp. 222–9, 225. Hessel Miedema, "The Term Emblema in Alciati," *Journal of the Warburg and Courtauld Institutes* 31 (1968), pp. 234–50.

30   A.D.S. Fowler, "Emblems of Temperance in the Faerie Queene Book II," *Review of English Studies* 11.42 (May, 1960), pp. 143–9, 144.

31   Christopher Wood, "The Perspective Treatise in Ruins: Lorenz Stoer, *Geometria et perspectiva*, 1567," in *The Treatise on Perspective, Published and Unpublished*, ed. Lyle Massey (Washington DC, 2003), pp. 235–58, 239. Raymond Klibansky, Erwin Panofsky

and Fritz Saxl, *Saturn and Melancholy: Studies in the History of Natural Philosophy, Religion and Art* (New York, NY, 1964). Lorenz Stoer, *Geometria et Perspectiva* (Nuremberg, 1567). The frontispiece includes two intersecting rotated cubes like Ryff's plinth.

32   Jamnitzer combined a method of perspective with "the study of elementary forces." Wenzel Jamnitzer, *Perspectiva Corporum Regularium* (Nuremberg, 1568), trans. D. Wade, *Fantastic Geometry: Polyhedra and the Artistic Imagination in the Renaissance* (London, 2012), pp. 88–9.

33   Giambattista Caporali, *Marco Vitruvio Pollione, De Architectura, Libri I–V* (Perugia, 1536), p. 32.

34   Margery Corbett, "The Architectural Title-page; an Attempt to Trace its Development from its Humanist Origins up to the Sixteenth and Seventeenth Centuries, the Heyday of the Complex Engraved Title-page," *Motif* 12 (1964), pp. 48–62, 58.

35   Near the frontispiece, Ryff describes the geometrical tools as "righteous." These tools have their basis in geometry that the artistic master builder cannot dispense with, so much so that "such instruments will be over-affectionated (*oberkünstlet*) from day to day, as you evidently will want to see in the figure set above." Rivius, *Vitruvius Teutsch*, Book I, p. 13. Translated by Ulrike Altenmüller.

36   Rivius, *Vitruvius Teutsch*, Introduction [n.p.]. Translated by Ulrike Altenmüller.

37   Ryff's title page of *Vitruvius Teutsch* identifies his book as for "all artful craftsmen, work masters, stone sculptors, master builders, well-makers, workers, painters, sculptors, goldsmiths, joiners and all those who might in the future artfully use the compass and straight edge for special purpose and versatile benefit … ." Rivius, *Vitruvius Teutsch*, title page. Translated by Ulrike Altenmüller.

38   Mircea Eliade, *The Forge and the Crucible: The Origins and Structures of Alchemy* (Chicago, IL, 1978), p. 29. See also: Klibansky, et al., *Saturn and Melancholy*.

39   Ryff uses different mottos with the frontispiece. The quoted motto appears in two of Ryff's sources: Tartaglia, *Nova Scientia* (1537), frontispiece; and Luca Pacioli, *De divina proportione, Facsimile of the ms. in the Bibliteca Ambrosiana di Milano (1458)* (Milan, 1982), "proverb," VIII r. Gerhard Harig, "Walter Hermann Ryff und Nicolo Tartaglia: Ein Beitrag zur Entwicklung der Dynamik im 16. Jahrhundert," *Physik und Renaissance; zwei Arbeiten zum Entstehen der klassischen Naturwissenschaften in Europa* (Leipzig, 1984), pp. 13–36, 15.

40   Rivius, *Vitruvius Teutsch*, Book I, p. 5. Translated by Ulrike Altenmüller.

41   *Melancholia I* also uses open and closed books. Klibansky, et al., *Saturn and Melancholy*.

42   Titus Burckhardt, *Alchemy: Science of the Cosmos, Science of the Soul*, trans. William Stoddart (Louisville, KY, 1997 [1960]), pp. 68–9, 201.

43   Gareth Roberts, *The Mirror of Alchemy: Alchemical Ideas and Images in Manuscripts and Books from Antiquity to the Seventeenth Century* (Toronto, ON, 1994), p. 107.

44   John Herman Randall, Jr., *The School of Padua and the Emergence of Modern Science* (Padua, 1961).

45   Pietro Pomponazzi, "On the Immortality of the Soul," in *The Renaissance Philosophy of Man*, ed. E. Cassirer, et al. (Chicago, IL, 1948), pp. 280–384.

46   In Ryff's discussion of the first book and chapter of Vitruvius, where the frontispiece is located, he describes "*Fabrica*" as "the work of the hand through which one may bring a thing into the work and other than that from certain cause … the origin of a thing may bring shape (Gestalt) into the work." Translated by Ulrike Altenmüller.

47   Rivius, *Vitruvius Teutsch*, Introduction [n.p.]. Translated by Ulrike Altenmüller.

48   Werner Oechslin, "Vitruvianismus in Deutschland," in *Architekt und Ingenieur: Baumeister in Krieg und Frieden* (Wolfenbüttel, 1984), pp. 53–76, 55. Harry Francis Mallgrave, "Introduction," in Mallgrave, Beasley, Baines, et al. (eds), *Mark J. Millard Architectural Collection*, pp. 8, 14–15.

49   Describing the human liver. Ryff, *Anatomy* (Strasburg, 1541), p. xxiiii, trans. K Crowther, *Adam and Eve in the Protestant Reformation* (Cambridge, 2010), p. 62.

50   Joseph Leo Koerner, *The Reformation of the Image* (Chicago, IL, 2004), pp. 39–41, 52.

51   Hermann Maué, "Nuremberg's Cityscape and Architecture," in *Gothic and Renaissance Art in Nuremberg, 1300–1550* (Munich, 1986), p. 46. Smith, *Nuremberg*, pp. 30–35.

52   Smith, *Nuremberg*, p. 36. This, despite classical architecture being associated with Catholic Italy.

53   Preserved Smith, *A Short History of Christian Theophagy* (Chicago, IL, 1922).

54   Koerner, *Reformation of the Image*, p. 311.

55   Peter Laven, *Daniele Barbaro, Patriarch Elect of Aquileia; with Special Reference to His Circle of Scholars and to His Literary Achievement* (University of London: Ph.D. dissertation, 1957).

56   Ralph Adams Cram, *Gold, Frankincense and Myrrh* (Boston, 1919).

57   Hans-Georg Gadamer, *The Relevance of the Beautiful*, trans. N. Walker (Cambridge, 1986 [1977]), p. 35.

58   W. Chandler Kirwin, *Powers Matchless: The Pontificate of Urban VIII, the Baldachin, and Gian Lorenzo Bernini* (New York, NY, 1997), p. 14.

59   Harry Francis Mallgrave, "Introduction," in Otto Wagner, *Modern Architecture: A Guidebook for His Students to This Field of Art* (Santa Monica, CA, 1988), p. 37.

60   Frank Lloyd Wright, "In the Cause of Architecture: The Meaning of Materials – Wood" (May, 1928), in Frederick Gutheim, *In the Cause of Architecture, Frank Lloyd Wright* (New York, NY, 1975), pp. 179–88.

61   See for example, Adolf Loos, "The Principle of Cladding," in *On Architecture*, trans. Michael Mitchell (Riverside, CA, 1995), pp. 42–7.

62   Koerner, *Reformation of the Image*, pp. 74, 411.

63   Cristina Ionescu, "Dialectical Method and Myth in Plato's Statesman," *Ancient Philosophy* 34 (2014), pp. 1–18.

64   Marco Frascari (email communication, June 6, 2005).

65   Paul Klee, *Notebooks, Volume 1: The Thinking Eye*, trans. R. Manheim (London, 1961), p. 407.

7

Displacing Matter

Manuela Antoniu

The title of this chapter alludes to the syntagma made famous by the celebrated anthropologist Mary Douglas, when she reprised the definition of dirt as matter out of place.[1] It seemed appropriate here to call on a culturally conditioned "undesirable" matter that, being diffuse, could be said to have existence but no specific form. For it is these valences—as conceptual categories—that make it analogous to the work described in the following pages: two projects executed in "condemned" buildings, that is, buildings whose demolition had been predetermined.

The title also brings in Georges Bataille's "declassification" of the category of form, and its consequent instability—grammatically oscillating between adjective and verb, thus escaping definition.[2]

Both projects were undertaken in Columbus, Ohio, USA in 1992 as collaborations between the author and American architect Michael Williams.[3] The first building we worked on had been erected, as a matter of urgency, on the campus of the Ohio State University in 1916 as a biplane hangar, where American students who volunteered for the front in the First World War were trained as pilots. We watched it being demolished to make room for a putative parking lot by a lone machine operator during the course of a single night in June 1992. The second building dated from the mid-nineteenth century, when it belonged to a sprawling campus for a seemingly thriving community of people then deemed mentally retarded.[4] It too had been purpose-built, accommodating a communal dining hall on the ground floor and a performance space on the upper level, where the residents held musical and theatrical productions for their own cultural entertainment. It was demolished in early 1993 for bafflingly vague socio-political reasons.

Our projects were opportunistic infiltrations into the space of action serendipitously opened up by the buildings' condemned status: on the one hand, the latter reified the buildings, relegating them to mere architectural objects soon to become waste matter; yet on the other hand it allowed us to test unprecedented architectural possibilities at a 1:1 scale, in a state of pure exception—the temporary suspension of any building regulations or building code.

In each case, our intervention was the result of an empathetic interrogation of the building, in the spirit of Louis Kahn's famously asking the brick "what it wants to be." This deferential questioning stance on our part was suggested particularly by the first building, which had been deliberately left in disrepair so that it might deteriorate to a degree that would render it unsafe; this was then construed as a retrofitted argument in favor of demolition as the only viable alternative. On both sites we worked from the premise that, faced with impending collapse, the buildings might wish to respond with a swan song, as it were. But how does a building gesture *in extremis*? Again, we took our inspiration from observing the extraordinary way in which the first building—whose roof's integrity (and therefore that of its truss structure) was being progressively compromised by daily rain—seemed to defy the ordinary logic of inanimate matter and instead behave as an assaulted organism would.

Before describing each project in more detail, some of the traits they shared could be summarized as follows: both were meditations on the imprint of function upon architectural space; both tested the capacity, embedded within each building's material substratum, for it to act upon itself by itself, with virtually no outer input; both rendered visible, through their respective materialities, the act of a building vacating its own space; and lastly, yet fundamentally, both projects posed the question of the extent to which architectural dematerialization is coincident with formlessness.

COMMUNICATIONS LABORATORY

Throughout the eight decades of its existence, the initial WWI biplane hangar was put to many uses, including a communications laboratory in the 1960s. At that time, a partial mezzanine was introduced into the vast space of the building's uninterrupted original volume, bisecting it horizontally. In order to accommodate wingspan (Figure 7.1), the structure had been conceived without supporting columns, its load being carried by deep rectangular timber trusses that doubled as clerestory fenestration. Working on the inserted partial floor, we used the building's potential energy to help it draw itself into itself by itself. This autologous movement was suggested by the improbable vectors of the building's material decay; for, in spite of being uniformly assailed by the elements, it degraded otherwise, as if retreating from both ends toward a healthy core that it wanted to protect, a center that indeed remained unaccountably intact.

The floor plan of the 1960s upper level included a series of small offices arrayed along one side of a linear corridor. Our intervention was to cut an identically positioned 7' × 8' module from seven consecutive facing walls belonging to these offices; to suspend the wall modules *in situ* from a steel beam that we had inserted and attached to the bottom cord of the trusses (Figure 7.2); to cut the floor area located roughly in the middle of the seven walls (having three on one side of it, four on the other) into two adjacent rectangular segments; to connect each of these, through a simple overhead system of crisscrossed cables and pulleys, to its

opposite far-end wall module (first and seventh, respectively); and to gradually load the two floor segments with wall fragments, measuring 1' × 8', that we had previously removed from the top of each contributing wall for the insertion of the steel beam.

7.1 The School of Military Aeronautics, 1918.

Configured in this way, and with the suspended modules ready to glide along the beam, the floor-and-wall ensemble responded to gravity's pull and movement ensued: the two progressively loaded segments descended through the floor as counterweights, causing the end wall modules to draw together (Figure 7.3). In turn, this had the effect of compacting the space of all intervening rooms and displacing them toward the middle, on either side of the cut in the floor.[5]

Thus a vertical motion of descent, coupled with a horizontal motion of compression, reduced the space of each room down to its functional elements. By virtue of this kinetic matrix, what used to be peripheral—a wall-mounted coat rack, a blackboard, a wooden staircase, a sink—gradually acquired centrality, eventually occupying the whole space bound by the room's parallel walls.

The example in Figure 7.4 illustrates this dialectic of inverse proportionality between the diminution of architectural space and the magnification of its erstwhile function. For not only does the sink now completely define the abbreviated room, but it does so at the very moment its function is annulled, that is, when using the

7.2   One of the wall modules in suspension.

7.3   Counterweights viewed from ground level, showing cable crisscross.

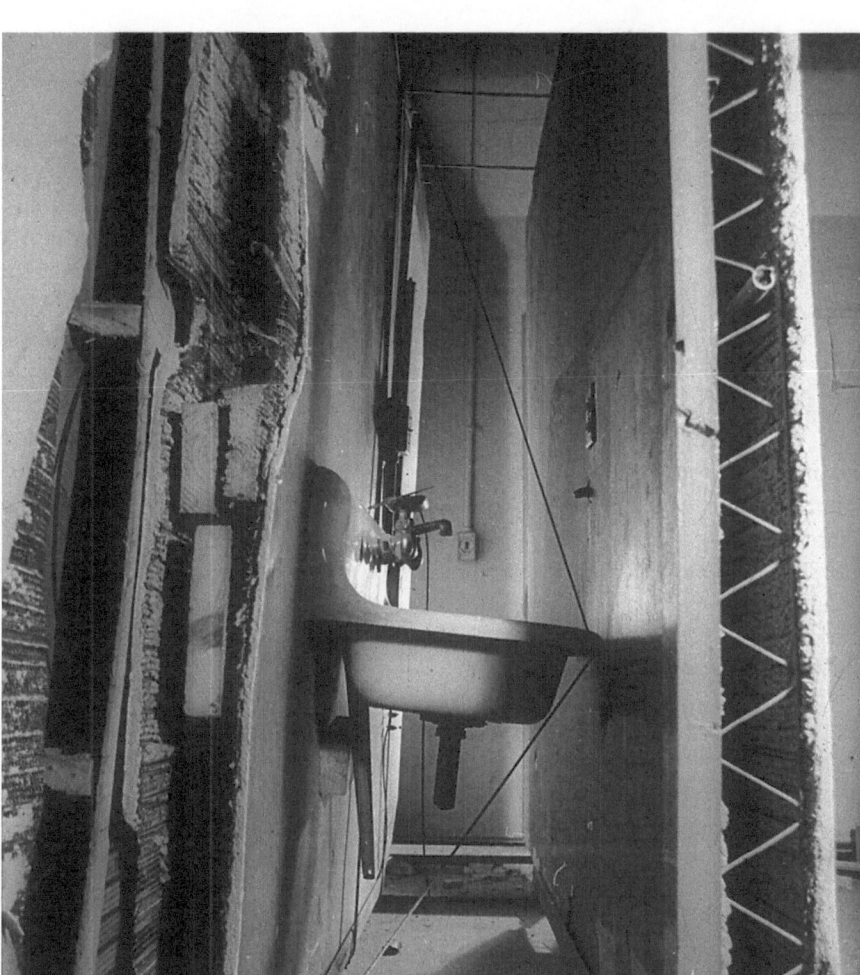

7.4   Walls 5 and 6 after motion.

sink has become a physical impossibility. This *reductio ad absurdum* helps stretch the notion of function past its figurative limit into the literal. This was all the more poignant in the case of two internally communicating offices, whose door opening was, after compression, closed by a wall.

If, in our intervention, the weft of the kinesis theatricalized the building's own retreat from the attack of the elements, the warp, on the other hand, weaved in the building's original function by addressing the gravity of belligerence. Fine archival footage exists of U.S. pilots during WWI, flying in formation over enemy lines that they subsequently bombed from biplanes similar to the one shown in Figure 7.1.[6]

If destruction from above (in the form of the elements) was visited on the building by the deliberate action of *inaction*, that is, by planned neglect, the building itself had fostered the visitation of violence from on high (in the form of manned bomber planes), which is a particular case of *un-doing* by an external agency. Although similarly charged with potential energy conferred by altitude, the counterweights, in contrast to the bombs, inverted the ground of undoing both by descending in

7.5   Truss view of counterweights and walls 1–4 after motion.

slow motion and by affecting the level above, rather than the one below them. If bombs were the lethal issue put forth by the belly opening of war planes, the cesarean section undergone by the floor of the communications laboratory showed, instead, how gravid with the possibility of self-birthing the building could be.

At the time of the building's inception, biplanes of the type shown in Figure 7.1 were turning aerial space into an unanchored, formless, boundless laboratory of experimentation, whereby initial observation missions subsequently morphed into bombardment raids.[7] Our experimental project in the tectonic laboratory intussuscepted the building's past with its imminent future, by defusing potential gravity from gravity's beckoning.

ONE DAY'S CHURNING

We undertook a similar attempt at barely assisted self-action on the part of the building in the second project, though this time not through the extravasation of the building's materiality; rather, through its introversion.

There, we put the former theater in the preparedness of staging its own—final—performance, as autogenously as its construction had occurred. For not only had the building been erected by the residents, but all the bricks that made up its fabric had been produced on site by them. Manufacturing their own building materials, tilling and cultivating the land, making their own clothing and shoes, producing

7.6   View from the stage.

their own food and entertainment spectacles, theirs was an entirely self-sufficient community. Among the archival photographs we perused before starting the project, all of them attesting to the former residents' industrious activities, we found one bearing the caption "One Day's Churning." It showed a worker in the dairy farm proudly gesturing to his day's labor: a neatly stacked pyramid of butter bricks.

The photograph became our own Louis Kahn brick: after electrically refurbishing the extant four banks of stage footlights, and supplying them with powerful lightbulbs, we suspended from the 9′ × 27′ proscenium arch a curtain made out of butter.[8] We then switched on the footlights and left them on for some two weeks. Throughout this interval, while the illumination made the curtain visible, the concomitant heat generated by the lights rendered it progressively invisible, by melting the butter in higher and higher registers and gradually transforming opaqueness into transparency. With this slowed-up curtain rise and its resultant unveiling of a mute dialogue between the empty stage and the long-absent audience, the theater performed its final act. But it not only performed it in the sense of being the sole active agent in a protracted one-act, one-actor dramatic play; it also *put on the performance* by drawing the curtain (however imperceptibly in real time), therefore signaling the start of the performance, all the while *performing spectatorship* to itself by itself.

Our registration of this process, using still photography at the rate of one shot per day, visually reads as a cinematic stop-frame sequence. The long-drawn-out, slow-motion progression that it

7.7   Curtain, day 2.

7.8   Curtain, day 9.

7.9   Close-up of curtain, day 14.

documents had isolated and magnified the theatricality of the encounter between an audience and a still-curtained stage. It had channeled, in the strength of the footlights, the collective gaze of anticipation—that ardent desire that the curtain disappear, in order for the performance to begin. And here the performance not only began, but all of it was played out in the dematerializing form of a stage curtain.

Although the complex political dialectics surrounding the theater building's history are beyond the scope of this chapter, suffice it to say that the project adumbrated them. As in the first project, this was expressed through an inversion: what had been produced in only one day during performative times in the life of the campus—as illustrated by the archival photograph of the proud dairy worker—was consumed on stage over a sequence of days when the specter of the building's demise loomed. However, both production and consumption revolved around the same material pivot, whose melting evoked elegiacally the tenuous nature of form.

DRAWING TO A CLOSE

Just as a building's "performance" encompasses not only a spectacular dimension but also an engineering one (we can speak of its thermal or energy performance, for instance), so the appellation "condemned" contains in its semantic register not only the designation of an architectural undesirable soon to be turned into architectural dust, but also an anthropic allusion. Condemned persons are marked by the inevitable lot in store for them. In the Renaissance, for example, only those who had been sentenced and executed served as human laboratories of anatomical exploration; and, given the link between the anatomical and the specular, their cadavers were moved around artists' studios by means of ropes and pulleys like so much puppetry, invested post mortem with improbable kinetic abilities.

Preoccupied, as we were, with the perithanatology[9] of architectural *dejecta*, our projects briefly turned both the communications laboratory and the theater into kinetic spaces. With minimal assistance from us, yet by virtue of their own materiality, they gestured with self-generated movements. In ontological terms, this poised them unstably between the animate and the inanimate. An anthology of resistance to destruction on the part of architectural structures is yet to be written, but disparate examples do exist, whereby matter presumed to be inert—and therefore unresponsive—has foiled attempts to obliterate it.[10]

Here, motion was not taken as an index of animism; rather, it brought into focus the dynamics of impermanence, of matter always in flux—the active interpenetration of form and formlessness *untethered* to a dialectics. Here, evanescent lines of motion traced non-material forms. And here, our visual registration of the process—the images having outlived the buildings in question—has recorded an absence as recursively as Bataille's entry "dictionnaire" and its precursor, Diderot and D'Alembert's entry "encyclopédie," in their respective eponymous works. We can thus say, with Bataille, that "for academics to be happy, the universe would have to take on form."

ACKNOWLEDGEMENTS

With thanks to Michael Williams and Frank Fantauzzi for securing access to the communications laboratory, this chapter is gratefully dedicated to Dr. Ella Chmielewska and Professor Mark Dorrian (University of Edinburgh), and to Professor Lily Chi (Cornell University) for her steadfast interest in the two projects.

BIBLIOGRAPHY

Aristotle, *Metaphysics*, trans. Hugh Tredennick, Loeb Classical Library (London: William Heinemann; Cambridge, MA: Harvard University Press, 1947): book IV. v. 1–vi. 6.

Bataille, Georges, "Critical Dictionary," trans. Dominic Faccini, *October* 60 (Spring 1992): pp. 25–31 (originally published in *Documents* 1/7, December 1929).

Douglas, Mary, *Purity and Danger: An Analysis of Concepts of Pollution and Taboo* (London and New York, 2009 [1966]).

Geist, Sidney, "Brancusi: The *Endless Column*" in *The Art Institute of Chicago Museum Studies* 16/1 (1990): pp. 70–87 and 95.

Haden-Guest, Anthony, "The Roving Eye," artnet. Available at: http://www.artnet.com/magazine/features/haden-guest12-29-99 [accessed November 23, 2013].

Wikipedia. Available at: http://en.wikipedia.org/wiki/File:Bombers_of_WW1.ogg [accessed December 31, 2013].

Wilkin, Bernard, *Aerial Warfare in World War One*. Available at: www.bl.uk [accessed December 31, 2013].

NOTES

1   "If we can abstract pathogenicity and hygiene from our notion of dirt, we are left with the old definition of dirt as matter out of place." *Purity and Danger*, p. 44. The "old definition" that Douglas refers to had been proposed by William James in his book, *The Variety of Religious Experience: A Study in Human Nature*, in 1929.

2   In his *Critical Dictionary*, under the entry "Formless," Bataille writes: "A dictionary would begin from the point at which it no longer rendered the meanings of words but rather their tasks. Thus formless is not only an adjective with a given meaning but a term which declassifies, generally requiring that each thing take on a form. That which it designates has no claim in any sense, and is always trampled upon like a spider or an earthworm. Indeed, for academics to be happy, the universe would have to take on form. The whole of philosophy has no other goal: to provide a frock coat for what *is*, a mathematical frock coat. To declare, on the contrary, that the universe is not like anything, and is simply *formless*, is tantamount to saying the universe is something like a spider or spittle."

3   Currently Assistant Professor in the School of Architecture at Louisiana Tech University, Ruston, LA (email: michaelw@latech.edu). Photo credits: unless otherwise specified, all Michael Williams except Figure 7.4, taken by the author.

4   These included women who balked at housework, for instance.

5   We had calculated for stasis between weight and counterweight to be reached when walls 4 and 5, facing each other across the opening in the floor, will have drawn quite close to its edges. The unequal depths at which the two counterweights reached stasis (as shown in Figure 7.3) are explained by the dissimilar materiality of the walls: some of them were plaster on metal lath, while others had a timber frame, thus weighing less.

6   In the public domain there exists a clip, lasting 1 minute and 12 seconds, from the Motion Picture Division of the U.S. National Archives, of a bombing run over German lines. It can be viewed online at: http://en.wikipedia.org/wiki/File:Bombers_of_WW1.ogg [accessed December 31, 2013].

7   See, for instance, Bernard Wilkin's synopsis, *Aerial Warfare in World War One*, on the new British Library site dedicated to the conflict. Available at: www.bl.uk [accessed December 31, 2013].

8   This was achieved by applying a uniform layer of butter to identical lengths of fine wire mesh, which we suspended from the proscenium arch. When concatenated, these produced the overall visual effect of a solid safety curtain (or fire curtain, more aptly).

9   I am indebted to Tsushimoto Sokun Roshi for the term, which he uses to denote the interval just prior, and immediately ulterior, to a person's death.

10   To give only two examples, Brancusi's Endless Column in Tirgu Jiu and the Palace of Culture and Science in Warsaw are well documented for having put up a struggle and won, albeit sustaining permanent scars in the process. See, for instance, Michal Murawski's work on the Palace of Culture; Sidney Geist, "Brancusi: The *Endless Column*" in *The Art Institute of Chicago Museum Studies*. See also Anthony Haden-Guest, "The Roving Eye," artnet. Available at: http://www.artnet.com/magazine/features/haden-guest/haden-guest12-29-99.asp [accessed September 2, 2014].

8

Jørn Utzon's Reverie of the Eye: Surfaces of the Sydney Opera House

John Roberts

Went to Yucatan. The ruins are wonderful so why worry? Sydney Opera House becomes a ruin one day.[1]

Jørn Utzon

INTRODUCTION: SOMETIMES DREAMS

During 2013, while construction works were being carried out at the Sydney Opera House, posters on site hoardings carried enthused messages and photographs from hundreds of people celebrating the Opera House and its architect Jørn Utzon (1918–2008). Australian and international visitors clearly hold the building and its architect in affectionate high regard; the rarity of such positive public emotion for architecture triggers the simple question of "why?" Historian Richard Weston regards the Sydney Opera House as "unquestionably the most popular, and arguably the greatest, public building of the twentieth century";[2] Japanese architect Tadao Ando regards the building as "the summit of the 20th Century Modernist architecture."[3] What might explain the aesthetic appeal of Utzon's Sydney Opera House, extending from the general public to historians and master architects?

The soaring white roof shells, the building's iconic and most visible elements, may hold a key. The tiled roofs, white like surf, cumulus clouds, sails, or paper can be seen from miles away; their surfaces can be touched by a small child; the roofs are visually sophisticated and materially complex, and are methodically resolved in a building comprising three sets of shells positioned on an artificial platform. The roofs are like things from a dream, an architectural "reverie," that is, a "moment or period of being lost, esp. pleasantly, in one's thoughts; a daydream."[4]

In Utzon's occasional articles or interviews the word "dream" has significance, especially if understood as "a visionary anticipation, reverie, castle-in-the-air";[5] and contemplation of the implications of Utzon's dream may offer a means to comprehending something of his methods. The appeal of Utzon's Sydney Opera

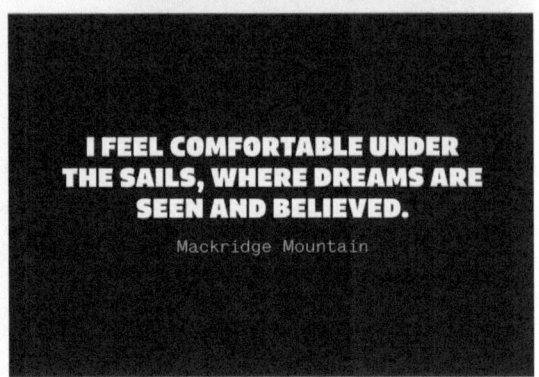

I FEEL COMFORTABLE UNDER
THE SAILS, WHERE DREAMS ARE
SEEN AND BELIEVED.

Mackridge Mountain

8.1 Poster detail, Sydney Opera House construction site, May 2013.

House, in particular its roofs, is considered in this chapter through a critical reverie on the roofs' material surfaces. The tiled surfaces appear to embody an architectural aesthetic of the eye, with visuality as a primary criterion for realizing the architect's dream of the building, itself seen as an expressive architectural reverie. Utzon acknowledged within his method the creative value of the dream, with the judgment of his eye and the democratic benefit of visitors as twin criteria, guided throughout by an "awareness that the building must provide the people who are to live in it with delight and inspiration."[6]

Critic Camille Paglia has put forward a theory of beauty based on the concept, deriving from Jung and Nietzsche, of the creative energy of Western culture's clear, youthful, shining "Apollonian" eye dominating the darker chthonian forces of "Dionysian" nature, of the bright sky rather than the heavy earth, an idea of a triumphant Apollonian ego which is "finite, articulated, visible."[7] Paglia's ideas, particularly her privileging of the eye as an organ of knowing, offer concepts to better understand the aesthetics of the built surfaces of Utzon's Opera House roofs. Seeing and the visual, however, are not widely trusted as reliable architectural means of knowledge or thought. Architectural phenomenologist Juhani Pallasmaa maintains that "All profound work arises from a dialogue between actuality and dream."[8] Yet Pallasmaa has repeatedly questioned the value of the instantaneous image, arguing theoretically for the epistemological authority of the haptic sense, promoting the tactile over the visual.[9] Yet Pallasmaa's argument, for the sustaining value of architecture's materiality against contemporary architecture's

8.2   Sydney Harbour and Fort Denison, east of Sydney Opera House.

transitory visual imagery and against "image products detached from existential depth and sincerity," is problematic, in that it seems to underestimate and undervalue architecture's visual power as it contributes to the poetic presence of architecture.[10] The visual architectural surface is considered here through a closer reading, avoiding hagiography, of the surfaces of Utzon's Sydney Opera House roofs.

Jørn Utzon reserved a particular place in his architectural ideals for dreams, as he noted on numerous occasions. His 1948 office manifesto "The Innermost Being of Architecture" concluded with the imperative that an architect "must have an ability to imagine and to create, an ability that is sometimes called fantasy, sometimes dreams"; Utzon saw the architect's capacity to dream alternatives to "the principle of what is most usual" as a necessary ability for creation of an architecture that would be "both varied and human."[11] Utzon expressed admiration for Louis Kahn's analogy of the university as a threshold between light and dark, where scientific knowledge, in Utzon's words, "fully exposed and totally exact, everything here proved and known," meets the realms of the dark and the unknown, of "ideals, dreams, aspirations, feelings, imagination, intuition"; from the dreams and feelings of the imagination, coupled with an exact and clear mastery of technology, "totally new things" can be created for people's delight.[12] To work towards building the new, to vary from "what is most usual," would require a dual confidence in the dream and the rational solution. The architect's eye, acting at thresholds of light and dark, would seem to be a most effective organ for making ultimate architectural judgments—especially at Sydney, on a naturally exposed site, on a culturally exposed building—on that which is most shared with the public eye, the outer surfaces of the large, prominent, public work of architecture.

SPIRITUAL SUPERSTRUCTURE

The Sydney Opera House offers people a reverie on architecture through experience of its material surfaces. Weston considers the roof covering of a million white ceramic tiles as "stunning, one of the most radiant and alive surfaces in architecture."[13] Amongst the shell roofs at the top of the Opera House's monumental stairs a bronze plaque displays a relief model of the roof forms with Utzon's words: "I arrived in October 1961 at the spherical solution shown here … it solves all the problems of construction."[14] A visitor can look up from the plaque at the white roofs' surfaces curving into the sky; these roofs are visible from infinite angles: from the city's approaches, from harbor waters, from an airplane above beaches and red-roofed suburbs. Utzon understood the roofs as vivacious visual sculptural forms, with surfaces to be seen and enjoyed unto infinity, as he wrote in *Zodiac* in early 1965:

> Looking at a Gothic church, you never get tired, you will never be finished with it—when you pass around it or see it against the sky. It is as if something new goes on all the time and it is so important … together with the sun, the light and the clouds it makes it a living thing.
>
> In order to express this liveliness, these roofs are covered with glazed tiles. When the sun shines, it gives an effect which varies in all these curved areas.
>
> We know it from these vigorous shapes as we actually experience them, we can see them as if we were sailing around the building.[15]

This is Utzon's visual manifesto. Eliding from Gothic to Modern, comprehending at once landscape and architecture, and visual and material appearances, it recalls the poetic methods of Swedish architect Erik Gunnar Asplund, for whom "the vault of heaven was the ceiling of the visual world."[16] This central poetic problem of transposing landscape and other concepts into architecture, between nature, image, and architectural abstraction, is approached and resolved throughout Utzon's Opera House designs; solutions to questions of spatiality and surface are evident in Asplund's ceilings in the Skandia Cinema and the Stockholm Library, and in Alvar Aalto's wave ceiling in the Viipuri Library, or the curved ceiling of the Seinäjoki Library. Sky, clouds, and waves were significant landscape elements for Asplund and Aalto, regarded by Utzon as his spiritual forefathers: "My family of architects goes back, and Asplund in Sweden and Aalto in Finland have something more than pure functionalism, they have sometimes what I would call spiritual superstructure. You call it poetry."[17] Utzon maintains inspirational continuity with his forebears as he affirms his methodical affinity with an essential idea of poetry.

SPHERICAL TRIANGLES

The sculptural roof forms of the Sydney Opera House are derived from the surfaces of a single sphere. The roofs are symmetrical along their ridges; their geometries

8.3   Jørn Utzon, Sydney Opera House, Main Hall roof.

are of forms and of surfaces, an array of varying spherical triangles produced, as Utzon explained, from the surfaces of a sphere.[18] Utzon assumed ultimate responsibility for the appearance of the building's surfaces, rendered through a roofing tile that might "produce the color, the surface texture, and the pattern required by me."[19] It is instructive and intriguing to compare Utzon's roofs with the white tiled outer walls of Arne Jacobsen's 1936 Texaco station at Skovshoved, as photographed by Utzon's friend Keld Helmer-Petersen in the 1940s.[20] Jacobsen's flat surface of vertically laid rectangular tiles on a small utility building contrasts formally with an attached toadstool-like canopy of smooth white painted concrete over the fuel bowsers. Utzon would have known this building, and it is tempting to imagine him considering and even working against the older Jacobsen's modernist/functionalist building while thinking about the Sydney roofs;[21] he exceeds Jacobsen's constructional scope and visual complexity with his vast tiled surface and its infinite edge curving against the sky, a three-dimensional outer surface turning from near vertical to near horizontal, both wall and roof, made of tiles in modular concrete elements, more like the earth's crust or a frozen sea than a building clad with tiles. Utzon was familiar with large developed surfaces from his childhood in the shipyards of Aalborg, and images of structures that vanish into infinitely large or small space—Mont-St-Michel, a Swiss mountain landscape, a dried sea, a sunset, a dark iceberg, a Chinese pagoda, algae, crystals—were included in the aesthetic manifesto "Tendenser i Nutidens Arkitektur" ("Trends in the Architecture of Today") assembled by Utzon and his friend architect Tobias Faber in 1947.[22] It appears that the ontological curve from finite wall to infinite roof at Sydney is presaged in these shining images of natural and built things tapering, rising, or vanishing into infinite distances.

FILM STAR

Many of the Faber/Utzon manifesto images display elements of Apollonian glamour, brilliantly challenging Pallasmaa's phenomenological ideal of "existential depth and sincerity" images of Hollywood by night, the palace of Versailles in a reflecting pool, rock crystals, sunbaking youths, sunset over an infinite sea, are contingent with a glamorous Hollywood image used to present Utzon's tall, glassy, prospect-dominant, even rakish, Langelinie project in 1953.[23] Architect Henning Larsen recalls that when an Australian journalist telephoned Utzon asking what he looked like Utzon replied, "Film star!"[24] Paglia traces images of shining glamorous surface from the radiant pagan pantheon of Olympus to Elizabethan armor to Hollywood publicity: "Movie stars of the Thirties and Forties, photographed in halos of shimmering light, had Spenserian glamour ... The camera gave them Apollonian power and perfection."[25] The kaleidoscopic geometries of 1930s Busby Berkeley musicals are replayed in the annual extravaganza of New Year's Eve fireworks over Sydney Harbour. A 1950s Swedish image of a swim-suited model in a torsional, Michelangelesque pose is part of the Utzon archive: the two-tone garment conceals the body's surface as it exploits and reveals its curves in its play of form, shadow, and radiant triangular surface pattern.[26]

"Glamour" means a kind of enchantment and magic, a notion arguably cognate with an architecture of surface; glamour is aesthetics, allure, danger, display; an appropriately revealing surface aesthetic for a house for the tempestuous, unreal rituals of the often immoderate art of opera. Vincent Scully saw the temples of Tikal as topographical architectural metaphors, "at once persons, mountains, and clouds

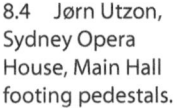

8.4 Jørn Utzon, Sydney Opera House, Main Hall footing pedestals.

... linking the Maya to the sun of life, as well as to the dark afterworld below."[27] Possibly when Utzon visited the ruins of Monte Alban and Tikal he imagined, beyond the realities of stairs and masonry, architectural beginnings in the reveries of the ritualized ancient cultures of Mesoamerica. Such associations, if not in Utzon's "dreams," are carried through poetic processes of metaphor into his architecture; there is a formal correspondence of appearance, a visual rhyme, between the feathered headdresses of Aztec and Mayan ceremony and the radiating centrifugal geometries of the Opera House roof chevrons, and in the extraordinary concrete footing pedestals that sit, massively radiant and flaring, beside the passageways inside the Major Hall.[28] Men wear feathers on their heads for combat or ceremonial panache; soldiers of the Australian Light Horse wore flamboyant emu plumes in their slouch hats.[29] Utzon, speaking of the shells, admits that the footing forms were a kind of "ornamentation" that "automatically emerged" as on leaf surfaces and in bamboo plants, describing the "natural, rhythmical ornamentation on their outer face."[30] The radiant circular geometries, echoing cultural glamour or derived from nature's visual poetries, are transmuted into the poetry of Utzon's architecture, enlightening the world of architecture and its discourses, possibly revealing undisclosed dimensions of architects and architecture.

THE RIGHT SOLUTION

The eye's seeing and the mind's dream of the Sydney Opera House intersect metaphorically in the white glazed tiled roof surfaces (Figure 8.5). Within Utzon's evolving "dream" of the building, from the competition entry of 1957 to the revised

8.5   Jørn Utzon, Sydney Opera House, restaurant roof.

scheme of late 1961, the resolution of the roofs' surfaces was key to the design and, ultimately, the successful construction of the shells. Chronologically, the roof tile study precedes the constructional solution: tile research took place during 1958–59, while Utzon arrived at the spherical solution in October 1961.[31] Utzon also noted his emphatically singular responsibility for judgment of tile color, texture, manufacture, and visual patterning in 1965: "The geometrical grids are therefore also influenced by my choice of materials, and the production method."[32] The visual, material, and constructional resolution of the roof surfaces predates and facilitates the structural and geometric resolution of the Opera House roofs, and is essential to their ultimate form.

Research and prototyping of roof tiles and tile fitting was carried out in Europe by Utzon and his Danish office, working with Swedish ceramicists Höganäs. The constantly altering curved roof shapes of the preliminary designs could not be satisfactorily clad with rectilinear tiles laid in either regular or random arrays; as a consequence of his researches, Utzon noted, "I therefore started to re-think the whole shape of the shells."[33] Utzon's spherical solution, using one sphere with a common radius for all the shells, with a geometrically regular relationship between square and irregular fragmented tiles over all the roofs, was the 1961 design breakthrough that ultimately enabled the Opera House to be constructed. Two types of tiles were used for the roof surfaces: square tiles, thick-glazed in white with a chamotte-enriched extruded ceramic body, were fitted as diamonds where the pattern required whole tiles; off-white matt tiles of varying shapes, trimmed according to their position, were set on eighteen different types of chevron-shaped panels pre-fabricated on site. Curved finely, almost imperceptibly in two directions, the "precast spheroidal lid elements" with their dual tiled patterns form the roofs' radiating spherical geometry, in both a constructional and visual sense.[34] Utzon emphasized the successful resolution of the tiling as key to solving the whole roof problem: "I think it is important to state that the struggle in finding the right solution for the pattern of the tiles was perhaps the main cause for this successful solution."[35] The tiled outer surface of the building, an ornamental reality and a visual reverie, ironically "underpins" the building's engineering solution, an inversion of normative architectural ontology, privileging appearance over construction, "visual decoration" over structure. The architect's dream and eye ultimately produced an answer to a fundamental architectural problem that had proved insuperable to Ove Arup's team of engineers working on the Opera House project from 1957 to 1961.[36]

AN EXCELLENT OBSERVER

Utzon outlined his architectural criteria in a 1983 article for his friend Denys Lasdun: "If you look at architecture ... evaluating a building purely from the sensation of joy it gives you—you experience the building with your senses only and you become a user of the building in the way the architect had conceived it,"[37] recognizing the importance of the visitor, the user, for whom the ultimate effects of the labor of architecture is intended. Weston, noting the timing of Utzon's 1948 manifesto calling for architects to use their capacity for "fantasy, sometimes dreams," relates how at the 1947 CIAM meeting Sigfried Giedion had appealed to architects to place art at the top of their agendas; Le Corbusier exclaimed in reply, "Poetry! The word must be pronounced! EMOTION, ART are as necessary as water and bread."[38] This new commitment, as Weston puts it, "to poetry, emotion, art, fantasy" pointed to a need for a freer, more sculpturally expressive architecture.[39] Giedion's words authorized a move towards more poetic discourses, a reverie of architecture in which Utzon was willing to take part.

One *OED* definition of reverie, "The fact or state of being lost in thought or daydreaming," is expressed in an 1802 usage that allies dream with observation: "The author of this strange and inconsistent reverie is, nevertheless, an excellent observer."[40] Architect Ole Schultz indicates Utzon's method of observation, recalling a sketch made on a cereal packet by the young Utzon while hunting with his father in Danish coastal wetlands:

> The sketch showed part of the Vejlerne landscape with ditches and fencing posts—drawn in perspective and with a small calculation at the bottom indicating that 40 fencing posts corresponded to 200 metres. It quite obviously caught Jørn Utzon's interest that it was easier to judge distances and find your bearings by allowing your eyes to wander along those lines in the landscape. There were also large drifting clouds in the sketch—with a clear dynamic direction determined by the west wind.[41]

Utzon's early method of knowing through observation uses regular fence-posts as tally-marks to calibrate landscape space, while noticing clouds and wind. His method of finding and using lines in the landscape to calculate landscape space in modules shows a logical mind behind an eye seemingly occupied with clouds. In Pi Michael's 1994 film *Skyer*, Utzon insists that his large-scale inspiration for resolving the Opera House geometry was not the orange skin as he had explained it in *Zodiac 14*, but was rather the surface of the inscribed globe of the earth.[42] Giedion also notes Utzon's description of this process: "As a result of my spherical system I can give all the dimensions their true size because I inscribe on the sphere great circles that intersect at the North Pole."[43] Utzon's metaphor is a trope at once Promethean and straightforwardly normative: a mode of thinking at a cosmic, planetary scale to arrive at a more straightforward way of giving dimensions "their true size" through comprehending an earth-scale geometry.

Utzon formed his "dream" of architecture as he traveled the world, taking movies and photographs, keeping few notebooks, writing in dog-eared picture books, accumulating images, and designing buildings.[44] He accumulated a bank of recollections at "true size," from recalled experiences of the natural world from landscapes to botanical details, and of ancient and modern works of architecture in Europe, Africa, Persia, Asia, the Americas. Utzon worked and travelled with his Nordic peers; he visited, corresponded, even worked for the master architects of his time: Aalto, Asplund, Mies van der Rohe, Wright, Le Corbusier. Utzon designed with full-scale prototypes and large models: for the Fredensborg housing project he worked with landscape architect Jørn Palle Schmidt, gauging wall openings, estimating landscape vistas, and visually estimating wall elevations on site, one by one, according to individual circumstance and from both inside and outside.[45] Utzon's architectural dream was unmediated, unreduced, consistent, and, it seems, experienced and remembered at full scale.

Utzon sought a perfect color and finish for the Sydney tiles through exhaustive investigation and full-scale prototype studies, including testing "ten different shades of white ... to determine the off-white color."[46] The tiles' dual surfaces recall living surfaces of textural and visual character: fresh matt snow and older wind-burnished snow; flat snow and shining ice; shining fingernails and matt skin.[47] An Utzon associate recalls an impromptu research excursion from a meeting in Zurich "because he felt forced to go up into the mountains in Arosa to study the white colors in the snow."[48] Architect Peter Myers recalls that Utzon gathered pictures in his office: he would "put together enlarged images from books ... And then he would just look at them, just as his hero Picasso would sit for hours in his studio ... ."[49] Art historian James Elkins endorses the eye's capacity to make accurate and memorable knowledge in this naively direct manner, "if we stop and take the time to look more carefully ... until the details of the world slowly reveal themselves."[50] A process of direct observation and memory of the world's details seems to have been central to Utzon's method; details of *surface*, such as on bright or windy water, on earth after rain, on ice and snow, as much as constructional architectonic details of jointing and flashing, require strong creative thought.

Utzon used a wave metaphor to visualize his thinking for the Opera House interiors in the journal *Zodiac* no. 14 in 1965, affirming the centrality of visual surfaces: "Decoration and color must be as organic a part of this complex as the white foam is part of the waves in order to achieve a complete and consistent character or style."[51] Through poetic reverie on the appearance of wave forms and foam, and drawing inspiration from both the natural world and from enduring human technologies, Utzon sought material solutions to the roofing problem in the enduring building materials of the ancient world, especially glazed ceramics.[52] Thus Utzon's thinking of wave images transmuted into the beautiful and enduring surfaces of half-glazed Japanese and Chinese ceramics, such as white-glazed Japanese Shino wares,[53] and the white glazed Chinese Song Dynasty (960–1127 CE) bowls described by potter Bernard Leach as simultaneously rough and smooth, like frozen snow's thawing surfaces.[54] A Song bowl once owned by Leach has vertical fluting in a carved lotus leaf design above a bare stoneware foot; the repetition of the stylized, geometric decoration and the depth of the white glazes recall the Opera House roof tiles fanning skywards from a bare concrete base.[55]

A HOUSE FOUNDED ON THE SEA

Poet and essayist Joseph Brodsky writes in the book *Watermark* of his delight in the surface effects of the light and water of Venice: "Surfaces—which is what the eye registers first—are often more telling than their contents."[56] Brodsky grants aesthetic primacy to his eye's reading and judgment of surfaces: "in the final analysis the eye is not so wrong, if only because the common purpose of everything here is to be *seen* ... the eye, our only raw, fishlike internal organ, indeed swims here."[57] Contrasting with Brodsky's wet Venetian eye is Camille Paglia's notion of the Apollonian eye of Western culture, an aesthetic of brightness, rationality, and measure, formed against the darker chthonian forces of Dionysian nature. Paglia privileges the eye as Western culture's principal organ of knowing: "The westerner knows by seeing. Perceptual relations are at the heart of our culture, and they have produced our titanic contributions to art. Walking in nature, we see, identify, name, *recognize*."[58] Paglia's aesthetic theory suggests that at the heart of Western culture is a "swerve" towards the sky, an aesthetic strategy "analogous to the shift from earth-cult to sky-cult."[59] Paglia finds the apotheosis of the eye in the linear aesthetic of ancient Egypt: "A black line on a white page. The Nile, cutting through the desert, was the first straight line in western culture ... Egyptian linearity cut the knot of nature: it was the eye shot forward into the far distance."[60] Paglia's figurative interpretation identifies the line of the river in the landscape, and recognizes and names the natural form as a precursor to a visually driven linear culture identified with light, distance, and shining surfaces: a culture at once triggered, formed, read, and enjoyed by the Apollonian eye. Utzon, in seeing, identifying, and recognizing the artistic value and significance of the shining surfaces of sea, snow, and ceramics, appears to proceed as Paglia indicates: in his architectural reverie, through sustained visual concentration, he transfers natural and cultural materials into architectural matter, of "rational" structural solutions and constructional strategies, clad in a kind of shining armor of "irrational," even ornamental, surfaces. Paglia argues for a material embodiment of sensory ideas, identifying poetry as "a sensory mode where ideas are or should be fully embodied in emotion or in the material world."[61] This approach recognizes Utzon's method as an overtly poetic synthesis, bridging between the material and the visual, a method that intertwines landscape, natural, and intercultural precedents to realize the Apollonian architectural surfaces of the Sydney Opera House roofs.

Literary critic Harold Bloom was a key influence on Paglia's methods; in a preface he cites Ralph Waldo Emerson's 1850 essay "Montaigne, or, The Skeptic." The strikingly odd metaphors of ships, shells, and "a house founded on the sea" uncannily evoke the architectural poetics, even images, of the Sydney Opera House:

> We want a ship in these billows we inhabit. An angular, dogmatic house would be rent to chips and splinters in this storm of many elements. No, it must be tight, and fit to the form of man, to live at all; as a shell must dictate the architecture of a house founded on the sea.[62]

The Opera House is visually and structurally "founded on the sea": Utzon describes it conceptually (even in Nordic terms) as "a house which is completely exposed," located "on a point sticking out into ... a very beautiful harbour, a fjord with a lot of inlets," recalling also the Nordic precedent of Denmark's Kronborg Castle, another large house on the sea.[63] Bloom notes the influence of the sea on the incarnation of poetry: "Poets tend to incarnate by the side of ocean, at least in vision, if inland far they be."[64] Utzon's work, with its affinities and proximities to the sea, could similarly be said to incarnate by the ocean. In the following section selected verses from *The Collected Poems of Wallace Stevens* (1954) are seen in light of Utzon's work, to consider coincidental poetic and architectural incarnations of materials of bright light, sea, sky, and houses by the sea.[65] It is intriguing to consider how the intuitions of Stevens and Utzon invoke and prompt feeling in visitors or readers, through poetic transformation of observation, insight, and feeling into works of considerable aesthetic power.

CELESTIAL PANTOMIME: COINCIDENCES OF UTZON AND STEVENS

Jørn Utzon and American poet Wallace Stevens (1879–1955) seem to share affinities in their work, particularly a theme of intertwining human feeling with the visual and sensual materials of landscape. Their poetic methods invoke diverse materials including natural phenomena of land, water, and sky; Stevens's title "Sea Surface Full of Clouds" evokes Utzon's well-known sketch of clouds and sea horizon. The poetic correspondences or coincidences considered here are interpretive and make no assumption that Utzon read Stevens. The purpose of this section is to suggest that literary poetry presents fresh insight into architecture, and that poetry itself as an expressive medium offers an expanded understanding of the poetic dimensions of architecture. Utzon's modes of articulation through building are, as he states, a result of both logical and poetic thought processes, in search of architectural "solutions" intended ultimately to delight and inspire people.

In Stevens's poem "Ploughing on Sunday" (1919) the narrator is outdoors on the Sabbath ploughing a paddock in the wind, turning over the soil of a field. A turkey-cock's white feathers and outspread tail toss and shine in an Apollonian display in the wind and sun. The poem is irreligious, pagan, and dazzlingly alight with the glittering outside world; even the heavy earth shines with wind-whipped water on this Sunday. The poem is built in short modules of utterance, like gasps of breath, shouts, or trumpet-blasts. The bright display of "Ploughing on Sunday" recalls the panache of the roofs of the Sydney Opera House on a bright windy day. Its metaphor of tilling a field as "ploughing North America" is a synecdochical leap of scale resembling Utzon's Promethean poetic feat of transforming great actual things—earth's curves, sea surfaces, icebergs, snowy mountains—into the modulated, geometrically realized surfaces of a work of architecture. "Ploughing on Sunday" illuminates Utzon's architectural methods, showing architecture as a world re-presented in new materials, forms, and scales; in return Utzon's architectural materiality shows how poetry co-opts and re-makes the world, re-animating it through metaphor, building new meanings through perceived similarities between things.

In Stevens's poem "Landscape with Boat" (1940) nothing happens, though much has happened already. A character is introduced, singular and contradictory, disciplined, uninterested in materialism though keen on flowery color, an "anti-matter-man, floribund ascetic." He is a rare character, perhaps an artist of large-scale works and ambitions, verging on heroic or mythic status: "He brushed away the thunder, then the clouds, / Then the colossal illusion of heaven. Yet still / The sky was blue. He wanted imperceptible air." The ambitions of this "floribund ascetic," a character reminiscent of Utzon, are celestial; he looks into an Apollonian world of thunder, clouds, heaven, blue sky, air, forming knowledge by seeing as well as through artists' dreams: "He wanted to see. He wanted the eye to see / And not be touched by blue. He wanted to know … ." This is creative eye-dominant Western visual culture in action: this ascetic artist's eye wishes to transcend even the limits of color in acquiring knowledge; the colossal scale of this vision recalls Utzon's synthesis of images, materials, measurement, technology, surfaces, and sustained research, the labor of mind and eye that was eventually needed to realize the Sydney shells.

Passages of the ten sections of Stevens' late poem "The Auroras of Autumn" (1947) dwell on themes of clouds, sea, and light. Although "Auroras" concerns farewell to life and to poetry, its scale and scope remain large: "Poetry stands here for all of life," claims Harold Bloom.[66] In "Auroras VI" Stevens likens the auroras of the title—the upper atmosphere's Aurora Borealis, which inspired Aalto's Finnish Pavilion for the 1939 New York World's Fair—to an ethereal theatre of light in the clouds, "a theatre floating through the clouds, / Itself a cloud, though of misted rock," where mountains are in flux, "running like water, wave on wave" in a dream-landscape of sky. An image of the "magnificence" of such "half-thought-of forms" recalls architectural dreams as in Utzon's early sketches for Sydney.[67] In "Auroras VIII" Stevens presents earth and the auroras as benevolent, truthful, innocent, harmless, "A saying out of a cloud, but innocence." Autumn's lights color the clouds but have no heat, no potential to burn or warm. They lack the strong light of "Ploughing on Sunday," the celestial will of the hero-character of "Landscape with Boat." Winter closes life's poem in ghostly solemnity, "this extremity, / this contrivance of the spectre of the spheres," a recollection of the sphere of the dreamed solution of the Opera House shells. Stevens ends "Auroras X" with a last image of light: "a blaze of summer straw, in winter's nick"; pale fire brightens one last time the niche of a fireplace or winter's imprisoning "nick." A long tapestry celebrating C.P.E. Bach's music decorates the low-ceilinged Utzon Room, a late Utzon design for piano recitals. This efflorescence of colored plumes is not in the sky but deeper in a "nick" in the building; Utzon saw the tapestry as a "visual experience" of music, a poetic translation "from the realm of the ear to the realm of the eye" once again, through poetry and into materials, feeling is crafted into a visible surface, as Utzon said, to "delight my spirit."[68]

CONCLUSION: REVERIE OF THE EYE

In Utzon and in Stevens, seeing begets feeling; through sight, the spectator, visitor, or reader's mind can align with the architect's or the poet's intuitive vision or reverie. Camille Paglia wrote in 2008 that poetry ought to directly address its public: "Poetry should not require academic translators to mediate between the poet and his or her audience."[69] Utzon's Opera House surfaces address visitor and spectator directly, from distant sighting to tactile encounter. Similarly the pages of Stevens's poetry address their complexities directly to the reader. Surfaces of words or of white ceramic tiles yield their significances to the eye directly and by degrees.

This chapter has looked at the surface of the Sydney Opera House roofs to argue that its visual surface was central to Utzon's design process, and that this centrality to development of a structural and constructional solution that might be both beautiful and practical, speaks for the high regard of the eye as it sees, recalls, and makes aesthetic judgment of architecture by its surfaces. Juhani Pallasmaa's skepticism towards the visual is suspended in the case of certain works of art, literature, and cinema; he acknowledges Utzon's architecture as "poetic alchemy [which] enriches the imagination of all of us."[70] Perhaps the surfaces of the Sydney Opera House roofs indicate the complex nature of knowledge, expressed in what can be seen, built, and written: the tension between the substances and surfaces of architecture; the fluidity of boundaries between the haptic and the visual, the cultural and the natural; and the ultimately accurate, even delighted, reverie of the eye.

BIBLIOGRAPHY

Bloom, Harold, *A Map of Misreading* (New York: Oxford University Press, 1975).

Brodsky, Joseph, *Watermark* (New York: Farrar, Straus & Giroux, 1992).

Caldenby, Claes, and Olaf Hultin, *Asplund* (New York: Rizzoli, 1985).

Clendinnen, Inga, *Aztecs: An Interpretation* (Cambridge: Cambridge University Press, 1991).

De Waal, Edmund, with Claudia Clare, *The Pot Book* (London and New York: Phaidon, 2011).

Elkins, James, *How to Use Your Eyes* (New York: Routledge, 2000).

Faber, Tobias, and Jørn Utzon, "Tendenser i Nutidens Arkitektur," *Arkitekten* 49 (1947): pp. 63–9.

Fromonot, Françoise, *Jørn Utzon: The Sydney Opera House* (Milan: Electa/Gingko, 1998).

Giedion, Sigfried, *A Decade of Contemporary Architecture* (Zurich: Editions Girsberger, 1954).

Giedion, Sigfried, *Space, Time and Architecture: The Growth of a New Tradition*, fifth edition (Cambridge, MA: Harvard University Press, 1967).

Helmer-Petersen, Keld, *122 Colour Photographs* (New York: Errata Editions, 2012 [1948]).

Keiding, Martin, Per Henrik Skou, Marianne Amundsen (eds), *Sketches: A Tribute to Jørn Utzon* (Copenhagen: The Danish Architectural Press, 2008).

Lasdun, Denys (ed.), *Architecture in an Age of Skepticism* (London: Heinemann, 1984).

Leach, Bernard, *A Potter's Book* (London: Faber and Faber, 1945).

Michael, Pi, *Skyer*, color film, 55 minutes (Copenhagen: DR–TV/The Danish Broadcasting Corporation, 1994).

Murphy, John, *The Studio of Jørn Utzon: Creating the Sydney Opera House*. Exhibition guide (Sydney: Historic Houses Trust, 2004).

Myers, Peter, "Joern Utzon," *The Journal of Architecture* 3 (Winter 1998): pp. 301–9.

Paglia, Camille, *Sexual Personae: Art and Decadence from Nefertiti to Emily Dickinson* (London and New Haven, CT: Yale University Press, 1990).

Paglia, Camille, "Final Cut: The Selection Process for *Break, Blow, Burn*." *Arion*, Third Series 16, No. 2 (2008): p. 22

Pallasmaa, Juhani, *The Eyes of the Skin: Architecture and the Senses*, 3rd edition (Chichester: Wiley, 2012).

Scully, Vincent, *Architecture: The Natural and the Manmade* (New York: St Martin's Press, 1991).

Stevens, Wallace, *The Collected Poems of Wallace Stevens* (London and Boston, MA: Faber and Faber, 1955/1954).

Utzon, Jørn, "The Sydney Opera House," *Zodiac* 14 (1965): pp. 48–9.

Utzon, Jørn, "Sydney Opera House: The Roof Tiles," *Architecture in Australia* (December 1965): pp. 85–92. Reprinted in Weston, *Utzon*, pp. 148–51.

Utzon, Jørn, *The Courtyard Houses: Jørn Utzon Logbook Volume I*, edited by Mogens Prip-Buus (Hellerup: Edition Bløndal, 2004).

Utzon, Jørn, *Two Houses on Majorca: Jørn Utzon Logbook Volume III*, edited by John Pardey (Hellerup: Edition Bløndal, 2004).

Utzon, Jørn, *Kuwait National Assembly / Prefab: Jørn Utzon Logbook Volume IV*, edited by Bjørge Nissen (Hellerup: Edition Bløndal, 2008).

Utzon, Jørn, *Additive Architecture: Jørn Utzon Logbook Volume V*, edited by Mogens Prip-Buus (Hellerup: Edition Bløndal, 2009).

Utzon, Jørn, Denys Lasdun, Norman Foster, "Royal Gold Medallist Jørn Utzon," *RIBAJ* (October 1978), pp. 425–7.

Weston, Richard, *Utzon: Inspiration Vision Architecture* (Hellerup: Editions Blondal, 2002).

Weston, Richard, *Materials, Form and Architecture* (London: Laurence King, 2003).

NOTES

1   Jørn Utzon, note to Bill Wheatland, 1966; in Françoise Fromonot, *Jørn Utzon: The Sydney Opera House* (Milan, 1998), p. 187.

2   Richard Weston, *Utzon: Inspiration Vision Architecture* (Hellerup, 2002), p. 114.

3   Tadao Ando, in *Sketches: A Tribute to Jørn Utzon*, eds. Martin Keiding, Per Henrik Skou, Marianne Amundsen (Copenhagen, 2008), p. 10.

4   s.v. Reverie, online *Oxford English Dictionary*, 3rd edn, March 2010. Available at: http://www.oed.com [accessed August 16, 2013].

5   s.v. Dream, online *Oxford English Dictionary*, 3rd edn, March 2010. Available at: http://www.oed.com [accessed September 9, 2013].

6   Jørn Utzon, in Weston, *Utzon*, p. 12.

7   Camille Paglia, *Sexual Personae: Art and Decadence from Nefertiti to Emily Dickinson* (New Haven, CT, 1990), pp. 5, 73.

8   Juhani Pallasmaa, in Keiding et al., *A Tribute*, p. 24.

9   Juhani Pallasmaa, *The Eyes of the Skin: Architecture and the Senses*, 3rd edn (Chichester, 2012).

10   Ibid., p. 33.

11   Jørn Utzon, "The Innermost Being of Architecture" (1948), in Weston, *Utzon*, p. 11.

12   Jørn Utzon, "The Importance of Architects," in *Architecture in an Age of Skepticism*, ed. Denys Lasdun (London, 1984), p. 214.

13   Richard Weston, *Materials, Form and Architecture* (London, 2003), p. 164.

14   Jørn Utzon, text on commemorative plaque at Sydney Opera House.

15   See Arup's summary of the developmental evolution of the roof solutions in Fromonot, *Jørn Utzon*, p. 84.

16   Elias Cornell, "The Sky as a Vault: Gunnar Asplund and the Articulation of Space," in Claes Caldenby and Olof Hultin, *Asplund* (New York, 1985), p. 30.

17   Jørn Utzon, Denys Lasdun, Norman Foster, "Royal Gold Medallist Jørn Utzon," *RIBAJ* (October 1978): p. 427.

18   Jørn Utzon, "The Sydney Opera House," *Zodiac*, no. 14 (1965): pp. 48–9.

19   Jørn Utzon, "Sydney Opera House: The Roof Tiles," *Architecture in Australia* (December 1965); reprinted in Weston, *Utzon*, p. 149.

20   Keld Helmer-Petersen, *122 Colour Photographs* (1948; New York, 2012).

21   Weston notes Tobias Faber's recollection that "as students their main enemy was the academic version of modernism as practiced by Arne Jacobsen and his contemporaries"; see Weston, *Utzon*, p. 18.

22   Tobias Faber and Jørn Utzon, "Tendenser i Nutidens Arkitektur," *Arkitekten* 49 (1947): pp. 63–9.

23   Weston, *Utzon*, p. 52.

24   Henning Larsen, in Keiding et al., *A Tribute*, p. 14.

25   Paglia, *Sexual Personae*, p. 177.

26   John Murphy, *The Studio of Jørn Utzon: Creating the Sydney Opera House* (Sydney, 2004).

27   Vincent Scully, *Architecture: The Natural and the Manmade* (New York, 1991), p. 16.

28   See cover of Inga Clendinnen, *Aztecs: An Interpretation* (Cambridge, UK, 1991).

29   Available at: http://www.awm.gov.au/encyclopedia/lhplumes/feathers/ [accessed August 16, 2013]. Note also Peter Aitken, "'Kangaroo Feathers' and the Mystique of the Light Horse," *Wartime* 14 (Winter 2001): pp. 50–53.

30   Weston, *Utzon*, p. 199.

31   Fromonot, *Jørn Utzon*, pp. 81–5.

32   Utzon, "The Sydney Opera House," p. 49.

33   Ibid.

34   Fromonot, *Jørn Utzon*, pp. 113–31; images and drawings are edited and reprinted from Utzon, "The Sydney Opera House," pp. 48–9.

35   Utzon, "Sydney Opera House: The Roof Tiles," reprinted in Weston, *Utzon*, p. 149.

36   Fromonot, *Jørn Utzon*, pp. 81–5.

37   Utzon, "The Importance of Architects," p. 214.

38   Sigfried Giedion, *A Decade of Contemporary Architecture* (Zurich, 1954), p. 42; cited in Weston, *Utzon*, p. 30.

39   Weston, *Utzon*, p. 30.

40   s.v. Reverie, online *Oxford English Dictionary*, 3rd edn, March 2010.

41   Ole Schulz, "Aage and Jørn Utzon Going Hunting," in Jørn Utzon, *Additive Architecture*: *Jørn Utzon Logbook Vol. V*, ed. Mogens Prip-Buus (Hellerup, DK, 2009), p. 4.

42   Pi Michael, *Skyer*, color film, 55 minutes, DR–TV, The Danish Broadcasting Corporation, 1994.

43   Sigfried Giedion, *Space, Time and Architecture: The Growth of a New Tradition*, fifth edn (Cambridge, MA, 1967), p. 681.

44   "Additive Explorer," in Utzon, *Additive Architecture*, pp. 230–312.

45   Jørn Utzon, *The Courtyard Houses: Jørn Utzon Logbook Vol. I*, ed. Mogens Prip-Buus (Hellerup, DK, 2004).

46   Utzon, "Sydney Opera House: The Roof Tiles," reprinted in Weston, *Utzon*, p. 149.

47   Weston, *Utzon*, pp. 152–3.

48   Børge Nissen, in Jørn Utzon, *Kuwait National Assembly / Prefab: Jørn Utzon Logbook Vol. IV*, ed. Børge Nissen (Hellerup, DK, 2008), p. 208.

49   Peter Myers, "Joern Utzon," *The Journal of Architecture* 3 (Winter 1998): 301.

50   James Elkins, Preface, *How to Use Your Eyes* (New York, 2000), p. vii.

51   Utzon, "The Sydney Opera House," pp. 48–9.

52   Utzon, "Sydney Opera House: The Roof Tiles," reprinted in Weston, *Utzon*, p. 148.

53   Peter Myers notes that scholar Else Glahn, who introduced Utzon to Chinese art and building, retained formulas for Shino and other ceramic glazes dating back to the Song dynasty, and passed these on to Japanese craftsmen in the 1930s. Peter Myers, in conversation with the author, October 25, 2013.

54   Bernard Leach, *A Potter's Book* (London, 1945), p. 242.

55   Northern Song Dynasty bowl, c.1025–1127. In Edmund de Waal with Claudia Clare, *The Pot Book* (London and New York, 2011), p. 258.

56   Joseph Brodsky, *Watermark* (New York, 1992), p. 27.

57   Ibid., p. 28.

58   Paglia, *Sexual Personae*, p. 15.

59   Ibid., p. 5.

60   Ibid., p. 59.

61   Camille Paglia, "Final Cut: The Selection Process for *Break, Blow, Burn*," *Arion* 16, no. 2 (Fall 2008): 22.

62   Ralph Waldo Emerson, "Montaigne, or, The Sceptic," from *Representative Men* (Boston, MA, 1850); cited in Harold Bloom, *A Map of Misreading* (New York, 1975), p. 177.

63   Utzon, "The Sydney Opera House," p. 48–9.

64   Bloom, *A Map of Misreading*, p. 13.

65   References throughout are to Wallace Stevens, *The Collected Poems of Wallace Stevens* (London and Boston, MA, 1955/1954).

66   Bloom, *A Map of Misreading*, pp. 187–8.

67   Jørn Utzon, preliminary sketches for Sydney Opera House, in Keiding et al., *A Tribute*, pp. 46–55.

68   Weston, *Utzon*, pp. 198–9.

69   Paglia, "Final Cut," p. 22.

70   Juhani Pallasmaa, in Keiding et al., *A Tribute*, p. 24.

9

On the Dissolution of the Modular Imagination

Dan Willis

If we decide to stop making things of the wrong size it will soon become clear that almost everything should be made quite a bit smaller. Things should only be big if they consist of a massing together of small units …[1]

Herman Hertzberger

THE PROBLEM WITH CONCRETE BLOCKS

A child's wooden block and the mason's traditional solid clay brick each have six sides. The rectilinear wooden block and brick can both be manipulated so that any one of the six sides is in the "facing up" position. Then the child or the mason can place another block or brick on top of that flat face. The exact position of the second "unit" need not be predetermined. The builder/child is free to experiment with the orientation of the second block, to make minute adjustments, or to introduce a cantilever (corbel). Even if the blocks or bricks are identical in size and material, the builder has a high degree of freedom, limited only by gravity and friction, in choosing where and how to place each successive unit. This mode of hands-on making, learned so early in our childhood, is "foundational" in both the literal and metaphorical senses of the word. It is playful, improvisational, tactile, another manifestation of the "dreaming hand" so eloquently described by Gaston Bachelard.[2]

It is through this basic act of piling one solid thing upon another that our toddler–masons learn the rules of infant architecture: all stable structures require a firm footing, lop-sided towers teeter and fall (just as do little boys and girls who lean too far), and a pyramidal construction that gets smaller as it grows upward tends to survive longer than one that gets bigger. Careful corbelling will allow them to deviate from the pyramidal tendency, but only within strict limits imposed by gravity. Any enclosed volumes will need to be created by some variation of vaulting. Young Frank Lloyd Wright learned these and similar principles using his solid wood Froebel blocks; Expressionist architects Bruno Taut and Hermann

Finsterlin explored the possibilities of building with miniature *Anker-Steinbaukasten* ("Anchor") blocks made of actual stone.[3]

Whether the blocks with which children play are made of wood, stone, or plastic, as long as the primary force that connects them is gravity, they will share an imaginative connection with earth, one of the four basic elements of the material imagination.[4] In all block buildings we will find vestiges of the two archetypes of earthen construction: the mountain and the cave. These correspond, respectively, to Bachelard's belief that the material imagination operates in two alternative modes: by elevating or by deepening.[5] Building upward, children challenge gravity and the limits of their own coordination. As Bachelard would tell us, their dreams of ziggurats, obelisks, and towers can be categorized among the "reveries of will." Alternatively, young builders may be inspired to seek stability, shelter, or concealment. Under the spell of deepening, they dream not of towers but of cellars, caves, and crypts—enjoying "reveries of repose."[6]

In contrast to the gravity-bound earthen imagination is the risk-averse mindset of the designer–builder in an industrialized society. This person's imagination occupies an abstract Cartesian space where height and depth have lost their ability to evoke either fear or triumph. The inevitable crashing to earth of the child's tower—which elicits a curious combination of regret and delight—is, of course, unacceptable when real human beings occupy a structure. Vital concerns for safety must be taken into account as we move from the playroom to the construction site. As licensed professionals, we architects can accept this responsibility to mitigate risk. The industrialized mind, however, is governed by priorities beyond mere safety. Efficiency and predictability are the chief concerns of the profit-motivated developer and contractor. A brick wall, and thus the bricks themselves, are not raw materials for an imaginative construction, but are conceptualized as components in a mechanized system. Each component part of the system—including the masons themselves—must be "optimized." The dreaming hand of the mason is disciplined, rapped on the knuckles with the ruler of Frank Bunker Gilbreth's scientific management, which reduced all possible manual movements to a combination of 17 basic motions.[7]

Gilbreth's systematized movements and processes similar to it are facets of the ongoing industrialization of building production. In previous writings, I have characterized this evolution as the transformation of work into labor, of craftwork into industrial production.[8] When the industrial mindset arrives at the jobsite and drafting room, the masonry unit is disciplined too. In contrast to the all-purpose solid brick, masonry units become specialized. Concrete blocks replace bricks for our utilitarian constructions and for most structural purposes. Being larger, partially hollow, and therefore quicker to lay, concrete masonry units (CMUs) allow walls and other structures to be constructed more efficiently than brick or stone. CMUs can also be shaped so as to allow the integration of steel reinforcing and concrete grout, transforming the masonry wall into a structural system that performs much the same as reinforced concrete.

There are other aspects to the specialization of CMUs: unique blocks are designed to serve as doorjambs and windowsills, as corners and copings, as lintels over openings, to absorb sound, or to form curved walls. As these technical refinements

are made, each type of block becomes progressively more adept at performing a specific task, and simultaneously less suitable for any other use.[9] This is not to say that a designer or mason cannot build imaginatively with the specialized CMUs. We need only recall Herman Hertzberger's design for the playground at the Montessori School in Delft, in which children fill the hollow cores of the concrete blocks with water and sand, to see that all physical objects can be re-imagined and re-appropriated in ways their manufacturer did not intend (Figure 9.1). As Hertzberger wrote, "perforated" blocks of this kind "look unfinished on their own and clamour to be put to some kind of use … they are an incentive to do something with them."[10] These improvisational re-imaginings are, however, creative *misuses* of the specialized block. They exist outside of the system to which the blocks belong.

9.1   Herman Hertzberger, children playing at Montessori School, Delft.

The creation of a system of industrialized construction presupposes that the creators of the system have already imagined nearly all possible applications. The designers of doorjamb CMUs need to have anticipated the characteristics of doorframes in a concrete block wall. Without a clear image of the finished product, it would be impossible to design the system components that will lead to the realization of that product. As Bachelard writes in *Water and Dreams*, "In the realm of aesthetics, this visualizing of the finished work leads naturally to the supremacy of formal imagination."[11] Much like a jigsaw puzzle, the components of such building systems are designed to fit together in a particular way. Improvisation is discouraged.

We can see a similar movement from the improvisational to the predetermined in the recent development of the popular toy plastic Lego bricks. The bricks are made by The Lego Group, a toy company based in Billund, Denmark that has been manufacturing plastic interlocking bricks in various colors since 1949.[12] The modern version of the interlocking brick, made of ABS plastic, was patented in 1958. The bricks are connected when "studs" on the top surface of one brick are friction fit between "tubes" on the bottom of another. This ingenious system allows the bricks to be offset or corbelled in single stud increments. The friction fit is just strong enough to hold the bricks together securely, but not so strong as to defy gravity, or to prevent children from easily taking the connected bricks apart. Original Lego bricks were two studs wide by four studs long. Just six of these original bricks can be combined in more than nine million possible ways.[13] According to Lego Group CEO Godtfred Kirk Christiansen, son of the company founder, the bricks were intended to be the basis of a system of play:

> Our idea is to create a toy that prepares the child for life, appeals to the imagination, and develops the creative urge and joy of creation that are the driving force in every human being.[14]

Originally, Lego bricks were sold in sets of various sizes, without detailed instructions. Although the company's catalogs and the boxes in which the sets were sold contained images of possible constructions, it was left to the children who played with the bricks to determine exactly what they would build.

Throughout the 1960s, Lego developed new variations of the basic bricks, and added features such as wheels and axles that made it possible to construct cars, trains, and other vehicles. By 1967, there were 218 different elements to the Lego system, including battery-powered electric motors. In the 1970s, Lego also introduced sets based on three themes: castles, towns, and outer space. The thematic sets, while still open to other interpretations and uses, directed children toward constructions in keeping with the themes. In 1974, Lego people, with round heads and moveable arms on brick bodies, began to be sold. More recently, Lego sets reproduce structures and vehicles from popular movies and cartoons, such as Star Wars, Spider Man, Harry Potter, Sponge Bob, and Toy Story (Figure 9.2). These sets are designed to produce a particular object: they include many specialized parts and very detailed instructions. There is one right way to build a Lego Star Wars battleship, and no room within the assembly instructions for improvisation or risk taking. By 2008 there were 4,200 different Lego brick shapes.[15]

A 2012 New York Times article on the evolution of Lego blocks characterized the situation this way:

> Even when actual bricks are involved, today's construction sets are often tied to billion-dollar franchises like "Star Wars" and the "Lord of the Rings"—and the story lines therein—and invite users to follow detailed directions, not construct their own creations from whole brick. It's less open-ended, some parents and researchers say, and more like paint-by-number.[16]

As one adult former Lego builder interviewed for the article put it, "what stinks about Lego sets now is that they're not imaginative at all."[17] More precisely, the makers of Lego bricks have determined that it is more profitable for them to de-emphasize their toy's ability to engage the material imaginations of children, and

9.2   Lego construction by Dara Hadighi.

instead to market the toys based on the cultural cachet of the finished objects children can make by following precise instructions and using highly specialized Lego components.

Absent the deliberate alignment with popular culture, there are many parallels between the evolution of Lego bricks and the development of masonry construction methods in industrialized societies. More specialized types of masonry have largely supplanted the all-purpose solid clay brick. Concrete masonry construction, as it has been developed in the United States, has displaced brick and stone from their traditional loadbearing roles. No longer generally used for structural purposes, brick and stone survive primarily as veneers. Cavity wall construction techniques consign the less attractive CMUs to the "Cinderella role" of hidden structure, while brick or stone, like the favored stepsisters, remain visible, kept around for their looks alone. Appearances, however, can always be simulated. If all we are after is "the look" of stone or brick, we can find more efficient ways to provide it. Brick veneer is therefore being replaced by thin bricks applied (or integrally cast into) precast concrete panels, and stone facings are being replaced by imitation stones made of concrete, which are attached to wall surfaces with adhesives. Anyone who has watched masons or tile setters glue pieces of simulated stone veneer to a vertical surface, starting at the top and working their way downward, has intuited an existential break with the "dynamic genius of reality while working with matter that resists and yields at the same time, like passionate and rebellious flesh."[18]

Just as the makers of Lego sets have substituted the prepackaged creativity of Lego's toy designers for the material imaginations of the children who build with their products, the industrial evolution of masonry materials into a construction system repositions the act of imagining a wall or structure from the mason's hand to the architect's or engineer's mouse and touchscreen. An embodied, gravity-bound, improvisational way of making is replaced by a mode of thinking that is primarily formal and technical. The apogee of this evolution is perhaps the replacement of human masons by robots, as seen, for example, in the brick constructions of architects Gramazio & Kohler.[19] This innovation acknowledges the reality that on the construction sites of twenty-first-century masonry buildings, the human masons are already working as if they were robots. Robotic masons are the logical extension of Gilbreth's bricklaying system. Were our human masons to work autonomously, if they built in ways that would take advantage of their skills and creativity, their services would not be as economical, as predictable, or as efficient as industrialized building production demands.

And yet, as long as we are piling up small things in order to make big things, a bit of the childish joy of building with blocks stubbornly remains.

DREAMING IN MODULES

> … as metre in music arranges the piece into segments thereby giving it lucidity, so the metric element in architecture makes distances and sizes intelligible. The size of objects is far more difficult to guess if they are flat and unarticulated than if they are divided up into units whose size is familiar to us, so that we can see the whole as the sum of its parts.[20]
>
> Herman Hertzberger

To Bachelard, a bit of wet clay or paste in one's hand was the epitome of the material imagination: "Paste is thus the basic component of materiality; the very notion of matter is, I think, closely bound up with it."[21] Alongside a dreaming right hand that kneads and molds a lump of clay, I propose another "basic component of materiality": a dreaming left hand that stacks, piles, and arranges fragments, blocks, or pieces of matter in order to make something larger. For most of human history the source of all things stackable was the earth itself. There is a kinship between the kneading and molding hand and the hand that stacks stones and bricks. Bricks are made of fired clay, and many thick-walled masonry structures, while technically the result of careful stacking, evoke images of spaces made by carving, digging, scraping, or other actions where material is removed. This gives to earthen structures a curious duality: while they may be the result of additive processes, imaginatively they can at the same time suggest subtraction. Peter Zumthor's *Therme Vals* (1996) project is perhaps the most widely known recent building that claims to be the result of surgically precise excavation and extraction.

The ability of most masonry materials to be readily shaped into concave or convex surfaces, or serpentine walls, suggests further ambiguity in terms of masonry's relationship to clay or paste. Jefferson's brick garden walls at the University of Virginia exhibit a materiality reminiscent of ribbons of homemade pasta. Many masonry buildings—the Baroque churches designed by Francesco Borromini, for example—appear to have been shaped by hand out of clay. Borromini seems to have been blessed with extraordinary patience: a mason–architect captivated by the sensuous possibilities of undulating surfaces realized through the laborious laying of individual stones and bricks.

The same masonry building may also exhibit duality in its gender: the interior of the cathedral is dark, cool, maternal, while the exterior pierces the sky with towers, pinnacles, and steeples. Earthen construction, of which masonry architecture is a subset, therefore satisfies the requirement for a fundamental substance of the material imagination. As Bachelard writes, "a matter to which the imagination cannot give a dual existence cannot play this psychological role of fundamental matter."[22]

At this point I wish to propose a second-order materiality, one step removed from "fundamental matter." This is the materiality of modular masonry, of building with objects that exhibit a degree of uniformity. Furthermore, I will call this mode of thinking and dreaming *the modular imagination*. The modular imagination is a hybrid. Although I have positioned it a step below the fundamental level of Bachelard's four elements, the modular imagination also serves as a bridge between the material and formal imaginations. In the mind and hand of true modular dreamer, the module works simultaneously in two modes: in both *combination* and *composition*. Thus, an architect who operates under the spell of modularity dreams of piling and stacking, the tangible acts of combinatory assembly, while at the same time imagining the unit as an ordering principle, as a way to divide, arrange, and define the space of a town, a building, or a room.

Unit masonry structures make not only distances and sizes intelligible, they also make visible the time of their own making. This impression is accentuated by the uniformity of the modular unit, which reflects the standardization in our common units of time: seconds, minutes, hours, days, weeks, etc. The walls of

9.3   Dan Hoffman, "Recording Wall," 1992.

brick and block buildings "tell tales" so that we may appreciate the work that went into creating them.[23] Architect Dan Hoffman's self-reflecting block wall project underscores (with intentional redundancy) this ability of masonry to mark time. Hoffman photographed himself as he built a wall of concrete blocks (Figure 9.3). To each block he applied a photograph of that exact block as he placed it in the wall. Hoffman's wall of blocks and photos revealed an open secret: the temporal awareness that is both obvious and yet usually not consciously noticed about unit masonry constructions.

Masonry constructions made of modular units also engage our imaginations because there is some ambiguity between the block wall under construction and the block wall in ruin. Absent the evidence of wear and weathering, or minus the false work and stockpiles of an active construction site, there is little to distinguish an unfinished wall from a crumbling one. This gives to all masonry buildings, and those made of modular units especially, an unfinished quality that is similar to a child's ongoing efforts to build with Lego blocks. The recently popular computer game Minecraft allows those who play it the ability to create and expand their domains indefinitely, using modular blocks of pixilated digital matter. Much of the attraction of the game, which is reported to have been played by over 40 million people, comes from its lack of a definitive end point or finish.[24]

The evocation of incompleteness, of a construction in the process of becoming (or disintegration) is a tell-tale sign of an active modular imagination. Marco Frascari identified the poured concrete ziggurat edges and transitions in Carlo Scarpa's Brion Vega Cemetery as the details responsible for the reading that the cemetery is a "built ruin."[25] The corbelled brick projections in the residential chimneys of Philadelphia architect Frank Furness, or the zig-zag courses of concrete blocks rotated 45 degrees in Mario Botta's house for Edmondo Pusterla, in Morbio Superiore (1982–1983), operate on a similar level. Unlike Scarpa's concrete ziggurats, the unit masonry constructions of Furness and Botta offer visible evidence of precisely calibrated time.

Botta's body of work is illuminating with regard to the hybrid nature of the modular imagination. In his early houses and in most of his smaller projects, Botta balances a playful exploratory attitude toward the masonry unit with a love of simple cubic forms. The dimensions and proportions of the houses are constrained by the module of the masonry units. From the exterior, the houses suggest solidity. Speaking of the Pusterla house, Botta stated that, "like a squirrel, I wanted to hollow out the form inside and cut the house out of the hill with a knife."[26] His playfulness is exhibited in his frequent use of corbels and the aforementioned horizontal courses of rotated blocks. Botta also liked to use monumental masonry arches, sometimes, as in his Barci House in Manno (1975) and Robbiani House in Massagno (1979–1981), as the single defining design element of the building. Botta's fondness for the "language" of unit masonry even extends to his designs for wooden furniture. For example, the central desk he designed for the main banking hall in the Banco del Gottardo, Lugano (1982–1988) exhibits the characteristics of horizontal masonry coursing, exaggerated by another favorite Botta device: the contrasting colors of alternating "courses."

In some of his larger commissions, we start to see the difficulty Botta has in translating the masonry dreams of his hands into the components of a major public building. For example, in his Ransila 1 Building in Lugano (1981–1985) we can see the familiar Botta masonry tropes: a cubic form that suggests a "hollowed out" solid mass of (in this instance) red brick, rotated brick quoins at the corners of the building, and brick corbels over the deep-set window openings. To underscore the image that the building is a piece of the earth, Botta even places a tree on its roof. The most dominating feature of the two street facades, what appear to be giant story-high corbels, are, however, corbel-like in shape alone. As a material expression, brick masonry corbels do not equate to story-high steel-framed cantilevers.

As the buildings he designed grew larger, Botta was forced to abandon the dreams of his hands stacking and carving the earth for the compositional preferences of his eyes. It is hard to maintain the illusion of carving into a solid block of the earth "with a knife" when a building occupies an entire city block. Floating panels of masonry supported by steel frames inevitably reveal their thinness. In a small building, economics may favor the traditional methods of local craftspeople; in very large buildings, it precludes them. The San Francisco Museum of Modern Art (1992–1995)—although technically clad in brick and formally in keeping with Botta's entire *oeuvre*—projects the materiality of a cardboard box. When Botta has the opportunity to return to small-scale projects, such as the ICF Showroom in New York (1985), the materiality of his modular masonry reappears.

9.4   O.M. Ungers, "Cube House."

An even more striking transition from a hybrid material–formal imagination to an attitude captivated by the formal properties of modules alone can be seen in the architectural career of Oswald Mathias Ungers (1926–2007). Some of Ungers's earliest masonry buildings, a student dormitory (1956), an apartment building (1958), and a multi-family house (1958–1959), all in Cologne, exhibit characteristics similar to Botta's small projects: deep-set openings, simple cubic massing, the impression of a shape carved out of solid earth. In the multi-family dwelling, Ungers even accentuates the brick corners with projecting "finger joints" (Figure 9.4). The brickwork is intentionally rough, even imprecise. The wedding cake massing, the incised openings, and the uneven brick surfaces together project an image of earthen solidity. Then, the heaviness of masonry disappears from Ungers's architecture almost entirely by the mid-1960s, making only a sporadic reappearance in 1989, with the design for the brick Bitz House, also in Cologne.

For Ungers this transformation of attitude was thoroughly intentional. Although he never lost his love of modular squares and cubes, as his career progressed Ungers grew to consider these and other geometric shapes solely from the standpoint of composition. "The essence of art and architecture is number, measure, and proportion. Sublimity is expressed through them alone."[27] As for materials, Ungers denied their relevance to his work:

> My concern is for pure form, for the abstraction, the basic type, the elementary building. No natural materials burdened with associations, no architectural details that tell the history of the window, the entrance, or the building with cornices, copings, and jambs, the differences between base and crown …
>
> Nothing metaphysical, no symbol for a condition of whatever sort. Only the direct, unadulterated form. Nothing moves, nothing is obscured, encoded or concealed …
>
> The aim is a non-objective architecture: space in itself, a building without qualities, encumbered by nothing, defined only by geometry and proportion, dependent on the rules of number and relation.[28]

In contrast to his projects from the 1950s, in Ungers's designs and—even more so—*drawings* from the 1980s and 1990s, one first recognizes the incessant use of the square as a planning module. The projects are presented as a series of nested square grids of various scales: room-sized grids within building-sized grids within urban grids. Trees are planted in grid patterns, the paving is based on a grid, and building elevations are gridded, as are the window mullions, the flooring, the light fixtures, and ventilation grilles. In these designs the disembodied square module pays no heed to gravity or orientation, as it slides effortlessly between scales, unencumbered with "associations" or "qualities."

> The new abstraction in architecture uses rational geometry with clear and regular forms in both plan and elevation. In this context, the plan is the result not of the literal interpretation of functions and constructive conditions but of logical geometrical systems.[29]

As with most architects who write about their work, Ungers's realized projects are less pure than his rhetoric suggests.[30] Nevertheless, with few exceptions, Ungers's architectural output was strikingly consistent over his long career. Francesco Dal Co called Ungers's architecture "a constant rejection of the allure of up-to-dateness."[31] Because his clients appear to have provided Ungers with sufficient and sometimes generous budgets, his built projects are generally made of high-quality materials. For this reason they never quite attain the weightlessness he extolls in his writings. The very precision and rigor of his buildings convey a sense of the effort that went into planning and building them, leading—ironically—to an awareness of their "embodied labor" not dissimilar from what Akos Moravanszky calls "the pathos of masonry."[32]

Naïve architects may assume that the Cartesian grid is nothing more than a neutral planning device, but in fact such instruments bring with them the rest of Cartesianism and its consequences: the separation of mind and body, the stoppage of time so that "nothing moves," the Benjaminian loss of aura, the exclusion of culture and tradition, the erasure of that which Ivan Illich called "the historicity of stuff."[33] To his credit, Ungers was not only fully aware of these consequences, he intended them. He crafted an ideology and a successful practice that was in many ways in perfect step with a global capitalist industrialized society.

The industrialist and the architect in the thrall of pure form share an aversion to the corporeality of things: the former because the real and the concrete are insufficiently predictable, the latter because they are impure. Neither views the materials and methods of construction as "an incentive to do something with them." Dreaming hands can be inefficient, and they have a tendency to deviate from "the rules of number and relation." Architects who manipulate abstract representations of pure form fit seamlessly into our contemporary conventions of building production:

> Efficient buildings are achieved in contemporary practice on the basis of a particular aspect of the construction process: architects who have designed in diagrams instruct construction supervisors in procedures to be implemented by builders. Influential in this restructuring has been another kind of efficiency, the economy of capital investment. Two results have followed: one, the reduction of the time of construction, shortening the interval between the project's inception and its potential occupancy; two, the following of construction procedures that use mass-produced parts and techniques developed in the modern movement, independent of its social and political aspirations. These procedures, along with the expectation of speedy efficiency, have affected architectural production at each of its stages.[34]

The citation is from Mostafavi and Leatherbarrow in *On Weathering*. Our contemporary need for "speedy efficiency" and our societal aversion to risk lead, they continue, to an ironic result: "mass production, which promises greater choice, has come in current practice to favor formulaic solutions."[35]

This is our current situation with respect to masonry construction. Concrete and brick unit masonry have evolved into a kit of parts from which the architect can pick and choose, leading, most often, to "formulaic solutions." The sorts of small masonry projects by Mario Botta, where the unity of the module as both a material and spatial unit was maintained, represent an approach to architectural production from another time. The speed of industrialized building suggests that we can plan everything in advance, whereas in the pre-modern era, significant buildings took so long to erect that changes to the social and physical environment around them demanded continuous adaptation and improvisation. Because we live in an age when "virtually everything concerning the building … is marked by the high velocity of late modernity and capitalism,"[36] the hybrid nature of the masonry module is being lost. It is this development that I have named "the dissolution of the modular imagination."

CONCLUSION: HERMAN HERTZBERGER, MASTER OF THE MATERIAL MODULE

> *The grid is like a hand operating on extremely simple principles … In its simplicity the grid is a more effective means of obtaining some form of regulation than many a finer-meshed system of rules which, although ostensibly more flexible and open, tend to suffocate the imaginative spirit.*[37]

> *Herman Hertzberger*

Comparing the published drawings for the design of Herman Hertzberger's best-known project, the Centraal Beheer Office Building (1967–1972), to drawings by O.M. Ungers for equivalently scaled projects, the initial visual impression one has is of striking similarity. They share a likeness in drafting technique (black ink line drawings on a white surface), a preference for axonometric projection to represent height and depth, and, in the building and site plans, an arrangement of spaces based on a modular grid of squares. Despite the reliance of both architects on grids as a planning tool, their writings and constructed buildings are evidence that their intentions could scarcely be more divergent. For Hertzberger, the grid is always incomplete. It is a scaffold, a framework, "which acquires its true identity by virtue of the those very interpretations that are given to it, notably the programmes that are filled in and the specific way in which this is done."[38] Rather than architecture ruled by "logical geometrical systems," Hertzberger believed that the best designs were the result of a reciprocal relationship between form and usage.

Hertzberger was influenced by the structuralism of Saussure and Lévi-Strauss,[39] and it is not much of a stretch to extend some of his ideas into the territory of architectural typology inhabited by Ungers. Yet, in practice, Hertzberger operationalized the idea of a "generative grammar," not through form alone, but through his favorite forms expressed in his favorite materials, notably hollow concrete blocks. These masonry units provoked the architect and the building inhabitants to "do something with them."

> *Such incentives are inherent in concrete perforated building blocks, representing as they do a basic and at the same time extreme example of the reciprocity of form and usage. The holes in these blocks just as literally demand filling in …*[40]

For Hertzberger, the module is always something tangible that has an inherent scale. Unlike Botta, Hertzberger intuited that the generative grammar of masonry could not be "scaled up" to the size of a building story. He discovered, however, that larger modules—city blocks or buildings—could be miniaturized, allowing one to create a miniature city within the interior space of a large building. (A general rule for all material things: made larger, they become more complicated; made smaller, they are simplified. Thus we live in a world where miniatures are common, but "maxiatures" are rare).[41] This is the secret to the success of Centraal Beheer. Despite a material palette typical of a low-budget parking garage, the building is a hospitable environment for office work. The relative proportions of the spaces Hertzberger intended as streets and those he meant as interiors are, certainly, a dominant factor in his having achieved these spatial readings. But there are other, equally important, factors: admitting natural light from above enhanced the illusion of an exterior space, while using "the kind of materials for floors and walls that we are accustomed to seeing out of doors" reinforced the impression further.[42]

Hertzberger's understanding of the module is also accepting of change over time. He argued against designing toward "an unequivocal goal," but instead advocated for an architecture that "still permits interpretation":

> What we make must constitute an offer, it must have the capacity to elicit, time and again, specific reactions befitting specific situations; so it must not be merely neutral and flexible—and hence non-specific—but it must possess that wider efficaciousness that we call polyvalence.[43]

In contrast to Miesian universal space, Hertzberger equips his spaces for unanticipated uses by positioning within them objects that invite our use. In his design for the Montessori School in Delft, Hertzberger includes in the hallways "podium-blocks" made of concrete and concrete masonry, which, at first, appear to be obstacles to both flexibility and movement. Yet the podium blocks operate just as the open cores of individual concrete blocks: they provoke the students to do something with them. They are used as seating, as gathering and meeting places, as outlooks, desks, or stages.

In the same building, Hertzberger designed a counterpoint to the raised podium-blocks. His description of the modular construction (Figure 9.5) in the kindergarten area has a distinct Bachelardian ring:

> The floor in the hall of the kindergarten section has a square depression in the middle which is filled with loose wood blocks. They can be taken out and placed around the square to form a self-contained seating arrangement. The blocks are constructed as low stools, which can easily be moved by the children all around the hall, or they can be piled to form a tower. The children also use them to make trains. In many respects the square is just the opposite of the [podium-block] in the other hall. Just as the block evokes images and associations with climbing a hill to get a better view, so the square hollow gives a feeling of seclusion, a retreat, and evokes associations of descending into a valley or hollow.[44]

Industrialized building production forces architects to thoroughly conceptualize a finished building in advance of its construction, a working method that "leads naturally to the supremacy of the formal imagination." Hertzberger's "polyvalent" constructions point to a way past the unimaginative uses of materials encouraged by contemporary practices. In Hertzberger's imagination the module is never stripped of its materiality. Even after his buildings are constructed his blocks, like a child's, remain tangible things that provoke playful experimentation.

9.5   Herman
Hertzberger,
recessed floor
with cubic wood
movable seats.

The hollow concrete block is the perfect metaphor for Hertzberger's approach to architecture. His buildings are humane places to live, work, and learn because they provide "openings" for their occupants' imaginations; openings "that demand filling in." As Bachelard warned in *The Poetics of Space*, buildings that are "final," that are too finished, too complete, do not invite our dreams.[45] It is telling that the published photos of Hertzberger's buildings, unlike those of Ungers (or most contemporary architects), always prominently feature people using the spaces he has designed.

Although Hertzberger is certainly well regarded in architectural circles, at least in the United States his achievements remain underappreciated.[46] This is unfortunate, because his work provides a model for the creative use of ordinary materials and common construction methods. His work demonstrates that the imaginative use of architectural materials need not be restricted to the buildings of the ultra-wealthy or to vernacular builders.

Hertzberger shows architects the virtue of buildings as "semi-finished products," which are able to address the demands of public safety, professional liability, and borrowed capital, yet still leave a few situations open to the dreams of future modular builders and dwellers. An active modular imagination gives us a middle road between designs where everything is predetermined and situations where all is left to chance. The material module is both a set of rules and an invitation to test the limits of those rules. "[T]he semi-finished product must consist of an inducement—and that is something which can only be achieved if it was your idea from the very start."[47]

BIBLIOGRAPHY

Bachelard, Gaston, *The Poetics of Space*, trans. Maria Jolas (Boston, MA: Beacon Press, 1969 edition).

Bachelard, Gaston, *Water and Dreams*, trans. E.R. Farrell (Dallas, TX: Dallas Institute, 1983).

Bachelard, Gaston, *Air and Dreams: An Essay on the Imagination of Movement*, trans. E.R. and C.F. Farrell (Dallas, TX: The Dallas Institute, 1988).

Bachelard, Gaston, *Earth and the Reveries of Will: An Essay on the Imagination of Matter*, trans. K. Haltman (Dallas, TX: The Dallas Institute, 2002).

Dal Co, Francesco, "Through Art, According to Will," in G. Crespi (ed.) *Oswald Mathias Ungers: Works and Projects 1991–1998* (Milan: Electa, 1998).

Frascari, Marco, "The Tell-the-Tale Detail," *VIA 7: the Building of Architecture* (Cambridge, MA: The MIT Press, 1984).

Gilbreth, Frank B. and Lillian M. Gilbreth, "Motion Study for a Crippled Soldier," in *Applied Motion Study: A Collection of Papers on the Efficient Method to Industrial Preparedness* (New York, NY, 1917).

Hertzberger, Herman, *Lessons for Students in Architecture*, trans. I. Rike (Rotterdam: Uitgeberij, 1991).

Illich, Ivan, H_2O and the Waters of Forgetfulness (Berkeley, CA: Heyday Books, 1985).

Lipkowitz, Daniel, *The Lego Book* (New York, NY: DK publishing, 2009; revised edition, 2012).

Moravanszky, Akos, "The Pathos of Masonry," in Andrea Deplazes (ed.), *Constructing Architecture, Materials Processes Structures: A Handbook* (Basel: Birkhäuser, 2006), pp. 23–31.

Mostafavi, Moshen and David Leatherbarrow, *On Weathering: The Life of Buildings in Time* (Cambridge, MA: The MIT Press, 1993).

Parkin, Simon, "A Journey to the End of the World (of Minecraft)," *The New Yorker* magazine, "Elements" blog: January 23, 2014. Available at: http://www.newyorker.com/online/blogs/elements/2014/01/a-journey-to-the-end-of-the-world-of-minecraft.html?mobify=0 [accessed August 30, 2014].

Pizzi, Emilio (ed.), *Mario Botta: The Complete Works, Volume 1: 1960–1985* (Zurich: Artemis, 1993).

Richel, Matt and Jessica McKinley, "Has Lego Sold Out?," *New York Times*, December 22, 2012.

Trachtenberg, Marvin, *Building-in-time: From Giotto to Alberti and Modern Oblivion* (New Haven, CT: Yale University Press, 2010).

Ungers, O. M., "Aphorisms on Architecture," in G. Crespi (ed.), *Oswald Mathias Ungers: Works and Projects 1991–1998* (Milan: Electa, 1998).

Willis, Dan, *The Emerald City and Other Essays on the Architectural Imagination* (New York, NY: Princeton Architectural Press, 1999).

NOTES

1   Herman Hertzberger, *Lessons for Students in Architecture*, trans. I. Rike (Rotterdam, 1991), p. 194.

2   Gaston Bachelard, *Water and Dreams*, trans. E.R. Farrell (Dallas TX, 1983), p. 13.

3   Information provided by Matthew Mindrup, based upon his doctoral dissertation; also via Wikipedia. Available at: http://en.wikipedia.org/wiki/Anchor_Stone_Blocks [accessed August 30, 2014].

4   Bachelard, *Water and Dreams*, p. 3.

5   Gaston Bachelard, *Air and Dreams: An Essay on the Imagination of Movement*, trans. E.R. and C.F. Farrell (Dallas TX, 1988), pp. 10–11.

6   Gaston Bachelard, *Earth and the Reveries of Will: An Essay on the Imagination of Matter*, trans. K. Haltman (Dallas TX, 2002), p. 7.

7   Frank B. Gilbreth and Lillian M. Gilbreth, "Motion Study for a Crippled Soldier," in *Applied Motion Study: A Collection of Papers on the Efficient Method to Industrial Preparedness* (New York NY, 1917), pp. 138–9.

8   Dan Willis, *The Emerald City and Other Essays on the Architectural Imagination* (New York NY, 1999), pp. 239–68.

9   The specialty blocks are not immune to misuse. Things can always be used in ways other than what their designers and manufacturers intend.

10   Hertzberger, p. 168.

11   Bachelard, *Water and Dreams*, p. 13.

12   Available at: http://en.wikipedia.org/wiki/Lego [accessed August 30, 2014].

13   Daniel Lipkowitz, *The Lego Book* (New York NY, 2009; revised edition, 2012), p. 35.

14   Ibid., p. 18.

15   Ibid., p. 36.

16   Matt Richel and Jessica McKinley, "Has Lego Sold Out?" *New York Times*, December 22, 2012.

17   Ibid.

18   Bachelard, *Water and Dreams*, p. 13.

19   Available at: http://www.gramaziokohler.com [accessed August 30, 2014].

20   Hertzberger, p. 200.

21   Bachelard, *Water and Dreams*, p. 13.

22   Bachelard, *Water and Dreams*, p. 11 (original text italicized).

23   The reference is to Marco Frascari's well-known writings on "The Tell-the-Tale Detail." According to Frascari's definition, every joint between masonry units is an architectural detail that "tells tales." Marco Frascari, "The Tell-the-Tale Detail," *VIA 7: the Building of Architecture* (Cambridge MA, 1984), pp. 22–37.

24   Simon Parkin, "A Journey to the End of the World (of Minecraft)," *The New Yorker* magazine, "Elements" blog: January 23, 2014. Available at: http://www.newyorker.com/online/blogs/elements/2014/01/a-journey-to-the-end-of-the-world-of-minecraft.html?mobify=0 [accessed: August 30, 2014].

25   Frascari, p. 35.

26   Mario Botta, cited by Emilio Pizzi, editor, *Mario Botta: The Complete Works, Volume 1: 1960–1985* (Zurich, 1993), p. 211.

27   O.M. Ungers, "Aphorisms on Architecture," in G. Crespi (ed.), *Oswald Mathias Ungers: Works and Projects 1991–1998* (Milan, 1998), p. 9.

28   Ibid., p. 11.

29   Ibid., p. 10.

30   How else to reconcile the man who wrote, "any building that has as its theme something other than itself is, from the spiritual point of view, trivial," with his design for a building that very literally resembles an ocean liner? The project is the Alfred Wegener Institute for Polar Research and Oceanography, Am Alten Hafen, Bremerhaven (1980–1984).

31   Francesco Dal Co, "Through Art, According to Will," in *Oswald Mathias Ungers: Works and Projects*, p. 6.

32   Akos Moravanszky, "The Pathos of Masonry," in Andrea Deplazes (ed.), *Constructing Architecture, Materials Processes Structures: A Handbook* (Basel, Switzerland, 2006), pp. 23–31.

33   Ivan Illich, *H_2O and the Waters of Forgetfulness* (Berkeley, CA, 1985), p. 3.

34   Moshen Mostafavi and David Leatherbarrow, *On Weathering: The Life of Buildings in Time* (Cambridge MA, 1993), p. 23.

35   Ibid.

36   Marvin Trachtenberg, *Building-in-Time: From Giotto to Alberti and Modern Oblivion* (New Haven, CT, 2010), p. xi.

37   Hertzberger, p. 125.

38   Ibid., p. 119.

39   Ibid., pp. 92–3.

40   Ibid., p. 168.

41   I owe my awareness of this general principle to Douglas Cooper, my former professor and drawing instructor at Carnegie Mellon University.

42   Hertzberger, p. 83.

43   Ibid., p. 152.

44   Ibid., p. 154.

45   Gaston Bachelard, *The Poetics of Space*, trans. Maria Jolas (Boston, MA, 1969), p. 61.

46   Hertzberger was awarded the RIBA Royal Gold Medal by the Royal Institute of British Architects in 2012.

47   Ibid., p. 164.

PART THREE
Imaginative Perceptions
of Architecture

10

Glass, as Light as Air, as Deep as Water

Ufuk Ersoy

Glass makes everything light. So use it on the site.[1]

Paul Scheerbart

Water, because of its intrinsic capacity to reflect, belongs to glass architecture; the two are almost inseparable …[2]

Paul Scheerbart

At the dawn of the twentieth century, glass had not yet gained a widespread use in architecture. For architects, it was still a mysterious substance that many of them hesitated to exploit. In the literature, however, it often and splendidly performed as an imaginary tool. Its perceptual perplexity inspired many fiction writers to question the visual definitions of everyday items.[3] It was the excellent metaphorical tool that could loosen the link between words and world by redefining surroundings in a completely different, that is to say, imaginary context. For instance, in the Bohemian poet Paul Scheerbart's visionary tales, glass was the unchanging constituent of ideal but unknown environments; it altered the face of the earth and made it appear in quite a different light. By eloquently deforming what one perceived, glass helped the writer to describe mysterious imaginative variations of landscapes that could not be mapped out in terms of objective knowledge, and thus could free the mind from ordinary images in memory. By virtue of its anamorphic quality, glass enabled Scheerbart to create puzzles of semantic dissonance open to interpretation and imagination. In fewer words, glass was the very substance of his imaginative freedom. It successfully contributed to his most significant literary tactic, "fictional estrangement."[4]

Scheerbart had no doubt that the imaginative aura of this substance would equally well serve to architects; particularly, to those seeking to challenge the nineteenth century's mechanized interiors and individuals alike.[5] To juxtapose architecture with glass would deviate buildings from their ordinary, pragmatic definitions. In 1914, Scheerbart presented his ideas to architects through his best-known *Glasarchitektur* (Glass Architecture) manifesto. The manifesto opened with a straightforward invitation to transform the environment by removing the closed character of the rooms in which modern men used to live.[6] In less than a decade, Scheerbart's provocative call to change the banal definition of architectural space by means of glass generated diverse and even contradictory results. Bruno Taut and Ludwig Mies van der Rohe were the two young architects who came up with two different seminal prototypes, which accurately mirrored how the architectural vision of the period moved back and forth between instrumental and communicative modes of understanding.

Writing *Glasarchitektur*, Scheerbart was already aware of Bruno Taut's engagement in the *Glashaus* (Glass House) Pavilion for the *Werkbund* (Working Federation) Exhibition at Cologne (1914).[7] As an "admirer" of Scheerbart's glass architecture fantasies, Taut designed the *Glashaus* to try out the architectonic capacity of the substance.[8] This experiment convinced Taut that glass was a dazzling artistic medium that could transform buildings into pure works of art and give architecture a "surplus" of meaning.[9] From *Glashaus* on, to persuade his colleagues to make use of glass turned out to be Taut's intellectual mission. During the First World War, despite his *"Glaspapa"* Scheerbart's harrowing early death in 1915, Taut did not lose his passion for glass and committed himself to represent the idea of glass architecture with his own words and drawings. His publications after the war, including *Alpine Architektur* (Alpine Architecture, 1919) were met with enthusiasm by a large group of young artists and architects who came together around Taut in the *Arbeitsrat für Kunst* (Workers' Council for Art) and *Gläserne Kette* (Crystal Chain) circles.

On the other hand, Mies did not share the same utopian vision with Taut and never participated in the circles Taut led. Mies's interest in glass developed later, upon his colleague Ludwig Hilberseimer's call for a more realist and practical implementation of Scheerbart's idea.[10] In 1920, having practiced for more than a decade in the offices of two recognized Berlin architects, Bruno Paul and Peter Behrens, Mies decided to reorient his view to find out the architecture embedded in the contemporary building techniques and materials. Distinctly, Mies was in pursuit of a language that would allow him to paraphrase the truth of architecture with more mature "artless words."[11] His entry for the Friedrichstrasse Skyscraper Competition in 1921 was the prelude indicating the transition in his architectural expression. Ironically, Mies's glass skyscraper proposal for the competition was eliminated for being an unachievable fantasy. All the same, Mies kept working on glass and developed his second glass skyscraper project *Hochhaus* (High Rise). The two glazed high-rise projects appeared in the fourth and last issue of the journal *Frühlicht* (Dawn) edited by Taut.[12] Mies, despite his disagreement with Taut's vision, knew well that *Frühlicht* was the only professional journal receptive to projects questioning the possible reality where he could publish his glass skyscraper projects.[13]

Taut's and Mies's expectations from the substance were relatively different. Yet, they shared the basic conviction that technology was the formative agent of the current culture, and they saw glass as the prospective element of modern architecture that they must of necessity experiment. Observed merely in consideration of this prevalent techno-centric stance, Taut's *Glashaus* and Mies's skyscrapers might seem to simply fetishize glass as the precious stone of the new industrial culture. In the two architects' eyes, however, far from being a garnish of the bourgeoisie, glass was the innovative substance that promised to redefine architectural spaces. In the words of their contemporary Arthur Korn, "[glass could] enclose and open spaces in more than one direction. Its peculiar advantage [was] in the diversity of the impressions it created."[14] Taut's and Mies's prototypes substantiated Korn's statement; they handled glass to open spaces in two different directions and to convey two distinct impressions. As will be revealed in what follows, while Taut wanted glass to soar as light as air, Mies sought to make it sparkle like deep, dark water.

GLASHAUS

Glass appealed to Taut because of its inconceivable fictive character that distinguished it from the rest of materials. Although it emanated from the earth, owing to its receptive nature dependent on ambient conditions external to it, it could act as if it were intangible. It could mimic "air, water, fire" in such a way that gravity might seem not to affect it. Thus, as the substance that triumphed over gravity, glass could appropriately let the architect elevate the material toward the immaterial. However, Taut's appraisal of glass was not limited to its capacity to narrate substantial transformations on the earth's surface. In addition, it could reveal some "subtle" values and feelings that would stimulate the human psyche; because, for Taut, overcoming gravity in building was analogous to overcoming one's body (or given conditions) in the world.[15] Concisely, Taut designed and built the *Glashaus* to set off two successive ethereal activities: aerial sublimation—the dynamism of air—that would result in psychological ascension—the dynamism of psyche.

The essay *"Eine Notwendigkeit"* (A Necessity), which Taut wrote right before the construction of the *Glashaus* began, clearly described the basic purpose of the pavilion. The contemporary architecture needed a *Volkshaus* (community building) that would reintroduce artistic spirit to people. Taut certainly borrowed this idea from his mentor at the Technical University of Munich, Theodor Fischer. Fischer intensely advocated that a new type of "house for all" that would bring all the arts together would spiritualize and reform the modern society.[16] Taut supported Fischer's reformist idea of a future art shelter for community. Yet, different from his mentor's progressive vision, the paradigm Taut envisioned for the *Volkshaus* derived from the past: the Gothic cathedral that had been idealized as the last example of authentic communal work since the early days of Romanticism. Precisely, in the Gothic cathedral, Taut recognized the act of "construction heightened to the point of passion" that he pursued. Even though it was built very simply and economically, it "transcended practicality" and converted "the most primitive form to a symbol."[17]

In 1912, the art historian Wilhelm Worringer had already historically justified the Gothic that Taut imagined.[18] For Worringer, the Gothic cathedral was the exceptional edifice that elevated architecture into art, and, therefore, deserved to be seen as the paradigm for a new northern art. More specifically, it embodied a *Weltgefühl* (world feeling) that moved the soul. The Gothic masters achieved this by dematerializing stone. Worringer advocated that "all expression to which Greek architecture attained was attained through the stone, by means of the stone," while on the other hand, "all expression to which Gothic architecture attained, was attained ... in spite of the stone."[19] The Gothic cathedrals let stone deny its essence, weight, and challenge the laws of gravity. Released from its weight, stone turned into a vehicle of "an immaterial expression," a bearer of "an uncontrolled upward movement."[20] In Worringer's view, the act of building was a struggle to awaken the dormant energies in materials. In this regard, in the Gothic cathedral, "there [were] no walls, no mass ... only a thousand separate energies speak to us."[21]

A pastel sketch Taut drafted in 1904, at the Collegiate Church of Stuttgart, represents how he read Gothic architecture (Figure 10.1). Sketching the nave of the church, Taut needed nothing more than the different tones of background color to articulate the structure. This technique gives the impression that a colored

10.1   Taut's sketch of Collegiate Church at Stuttgart, 1904.

glow tinges the vault and lightens the solidity of the structure by making stone, the symbol of gravity, disappear. As a result, while the arcade becomes hardly discernable, the converging vault ribs soar even higher above the nave-like latticework. The diaphanous surface denies the material limits of construction but still succeeds in setting spatial limits. Similarly, in the *Glashaus*, Taut designed the colored prismatic glass wall wrapping its prominent dome to eliminate any feeling of heaviness by transforming the solid surfaces of the structure into a weightless layer of colored light.

Under the sunlight, the cupola of the glass pavilion seemed to dissolve into a spectrum of colors, which developed from a deep blue, to a moss green and then to a golden yellow. Their culmination at the peak was a dazzling creamy white. Glass walls, which refracted the daylight in various hues, created a dynamic surface that apparently absorbed the reinforced concrete ribs carrying the glass panes. The thick diaphanous glass walls did not allow for any visual contact with the outside world. Likewise, there was not any horizontal reference in the hall that would direct the eyes to the cupola's connection with the ground. In the *Glashaus*, through colorful luminous surfaces in constant flux, Taut created perceptual ambiguity, which encouraged the impression of architecture without earthly limitations that must have agitated visitors' sense of horizon. Thus, after visitors entered the glass hall, their contact with mundane reality was hypothetically suspended.

From a distance, the *Glashaus* looked like a gemstone that grew up from the ground. Coming closer, one would notice a substantial transformation from the bottom upwards. The building, which emerged from an organically shaped concrete base, ended in the crystalline geometry of the polyhedral cupola, which was composed of rhomboids. The transformation from soil to crystal, from base to dome, hinted at the relationship between the two basic parts of the building, which aimed to mirror the structure of the universe, while also analogizing Taut's imagined transformation of the mundane human body into a transcendent spirit. Taut's program for the structure was to integrate the dome, which characterized the celestial order, with the hidden cave-like interior housed within the concrete base, which symbolized the fecundity of the Earth. Similar to the Gothic Cathedral, Taut's building program was based on a hierarchy of light and aimed to establish continuity between the inner and outer world. Ultimately, it was a creative microcosmic reproduction that sought to remind people of the correspondence between architecture and universe, a renewed awareness that would result in a self-crystallization. Simply put, in Taut's own words, "the act of construction in a more elevated sense," could draw one to an aesthetic experience that would reactivate "the faculties of the soul hidden behind the veil of faith."[22]

The art critic Adolf Behne, who hoped for a renewal of arts, which would generate a spiritual revolution and change the modern individual, appreciated the *Glashaus* having only an "inner-artistic" purpose.[23] Taut's building was a purpose-free—*zweckfrei*—artwork that freed technology from its pragmatic concerns. To achieve this, Taut returned to "the primal elements of building," the wall and the opening, and exposed them to the "reality of arts." The primal elements were enlivened through "pure" artistic means of color, line and light. Leaving aside all derivative elements, Taut invented his own modern ornament that was stripped of all excess and was purely expressive. In Behne's view, Taut subdued construction techniques to artistic expression and converted the *Baukunst* (art of building) into a primitive, cosmic Ur-force, capable of transforming the world. It provided the concrete model of a new kind of architecture closer to crystalline, abstract, non-historical forms, distinct from the pseudo-symbolic buildings of technology, such as Behrens's well-known AEG Turbine Factory (1908–1909). For Behne, *Glashaus* stood as a prototype that allowed him to reconcile artistic creativity with matter-of-factness and construe the principle of *Sachlichkeit* (objectivity) as a synthesis of reason and vision—*sachliche Kunst*. Mies, however, read the same principle, *Sachlichkeit*, from a completely different standpoint.[24]

GLASS SKYSCRAPERS

In 1919, the rejection Mies received for the *Ausstellung für unbekannte Architekten* (Exhibition of Unknown Architects), which gathered projects offering "radical solutions to [contemporary] problems," provoked him to reconsider his own design strategies.[25] He became increasingly engaged in theoretical discussions questioning the fundamentals of artistic *Gestaltung* (form-giving), with the hope of more fully understanding what constituted architecture as a work of art. The same year, after the exhibition closed its doors and Taut's *Alpine Architecture* was already in the hands of its readers, Ludwig Hilberseimer warned architects against prevailing misapplications of Scheerbart's glass architecture. According to Hilberseimer, Taut and his circle misread Scheerbart by closing their eyes to the existing conditions. The essential "constructive premises" of architecture were sacrificed for the sake of fantasies.[26] For Hilberseimer, Taut's speculations of achieving a faultless world and society simply by adorning the Alps with crystalline glass buildings denoted nothing other than an escapist imagination that ignored the difficulty of offering concrete alternatives to the present conditions. Years later, Taut regretfully admitted that the freedom of discretion on paper led him to easily overlook the given reality of the material world in which he lived, and ended in an unachievable abstraction.[27]

Hilberseimer's criticism was an open call for the projection of a more realist and practical glass architecture based on the principle of *Sachlichkeit*. Mies took this call on as his mission in the Friedrichstrasse Competition.[28] The demand for *Sachlichkeit* asserted, for him, the end of art in architecture. But still, pure expression was the identical goal common to Mies and Taut. To attain an elementary but more realist language, Mies put his focus on building technology. Explicitly, industrial materials and construction techniques came before the pure artistic means as a precondition for architectural design. The basic concern was to conceive the primitive and eternal Ur-form immanent in modern technology even if it had been intentionally ignored or masked out by the previous generation. In the first paragraph of his short article in *Frühlicht*, which broke his long reticence in architectural media, Mies clarified that "the bold constructive thoughts" that give skyscrapers their strong impression were only visible on the ones "under construction."[29] Traditional walls concealing steel skeletons completely killed this impression.

Mies was certain that the latest non-load-bearing glass walls would better represent the constructive principle of the new age. Even so, he did not forget that glass itself required a unique formal approach pertinent to its own characteristics. Otherwise, this time, large glass panes hung on the steel frame would kill the façade. "To avoid the danger of lifelessness," in his proposal for the Friedrichstrasse competition, Mies decided for a mass composed of three polygonal towers. None of the "façade fronts" of the polygonal structure was parallel to the peripheries of its triangular site. He was keen on these angled glazed façades, because "it was not an effect of light and shadow [he wanted] to achieve but a rich interplay of light reflections."[30] For Mies, the "man of few words," imagination cannot be explained by writing.[31] It is of no surprise, then, that he left unexplained what reflections on glass surface brought in his architecture and what distinguished his approach from that of Taut and others. Nevertheless, the recognized perspective of Friedrichstrasse Skyscraper tells more (Figure 10.2).

Although the large-format image rendered in charcoal initially brings to mind the technique of chiaroscuro, it derived from a series of photomontages.[32] Mies

10.2   Mies van der Rohe, Friedrichstrasse Skyscraper Project, 1921.

consciously decided on the technique of montage to experiment how his glass skyscraper would look in its urban context. At the outset, he inserted a rough outline of his building into the enlarged eye-level photograph looking north along the Friedrichstrasse. Then, he cleaned the scene from all distractions of the hysterical metropolitan life by darkening the surroundings of his drawing. Finally, to expose the skyscraper, he cropped the frame and meticulously articulated the reflections on its glazed surface. Later in his career, Mies frequently used the same technique to control and manipulate the settings and optic qualities of his building photographs in a way that denied the reality.[33] However, in the Friedrichstrasse project his intention was not to create a photorealistic simulation. By superimposing, dissolving and cropping, he abstracted the existing conditions and located his building at the Friedrichstrasse that he dreamed. He made use of the technique, as the filmmaker Lev Kuleshov suggested, "to create a new earthly terrain that did not exist anywhere,"[34] whereas his technique to articulate the presence of glass making up the façade was chiaroscuro.

Needless to state, on the image, the glass skyscraper stands out as the most active figure while the city mutely recesses into the background. It is the only building that communicates with its milieu. Its glazed façade fronts at once reflect the sunlight and pick up on the shadows of surrounding buildings together with those of the interior. Though dark shadows on the surface give clues about vertical volumetric recesses and the slabs behind the glass, luminous reflections that Mies highlighted with white prevent the eyes from focusing inwards. In contrast to the sharp edges, which precisely define the skyscraper's geometry, the glazed surface remains elusive. This perceptual instability imbues the glazed surface and the building with a sense of depth and mystery. As the term "reflections" implies, Mies's intent was to liquefy the surface by means of glass. He exploited the technique of chiaroscuro to emphasize the receptive character glass similar to water. For the modern eye, tired of looking at congested opaque walls, the glass wall being like a liquid surface, was a cure. It could give depth to pure vision by urging sensual limits.

The polygonal shape of Mies's twenty-floor story glass building made critics like Carl Gottfried think of a "tower-like, Gothic force," and compare it with Crystal Chain member Hans Scharoun's watercolor crystalline tower designs.[35] The glass skyscraper indeed aimed to embody the crown of the city (*Stadtkrone*) that Taut advocated. Yet, Mies was interested neither in colored light nor in formal distortions that challenged the perspective. For Mies, glass was cold and colorless. Similar to the walls of the *Glashaus*, the glazed façades of the skyscraper were in flux. What set the façades in motion, however, was not the colored prismatic structure of glass that transformed sunlight into a shadowless colored light but the reflective patina that liquefied the surface and made it act like water. Mies's scrutiny of reflective liquefied surfaces reached its peak at his Barcelona Pavilion, which marked the postlude of Mies's transition period that had started with the Friedrichstrasse project.

Similar to the glazed surface of the skyscraper, reflective surfaces in the Barcelona Pavilion did not allow for any clear vision. Along with the labyrinthine layout of the building, polished glass and marble surfaces transmuted the rational structure into a perceptually ambiguous space. In *Glass in Architecture and Decoration* (1936), Raymond McGrath maintains that Mies polished glass as "the modern counterpart, on a larger scale, of the Claude Lorraine glasses," to create a stereoscopic effect.[36] Similar to the Lorraine glasses in the eighteenth century that transformed surrounding landscapes into a distant picturesque image, polished surfaces of Mies's pavilion visually dislocated the things in and around it. Superimposed reflections on the marble and glass surfaces, relocated the things on a different, fictive landscape. Robin Evans described his experience in the reconstructed pavilion as a "dreamy disorientation."[37] Mies made use of reflective surfaces as his anamorphic tools that simultaneously created an aesthetic distance between the pavilion and its observers, and intensified the feeling of horizon by visually extending the platforms and walls defining the boundaries of the building. A polished, reflective surface was a dream device that opened Mies's building to imaginative speculations of its visitors and their self-discovery. But, unlike the hermetic microcosm in Taut's *Glashaus*, as Detlef Mertins observed, the self-discovery the Barcelona Pavilion promoted was not only from within but in relation to the milieu.[38]

MATERIAL CAUSE

As a substance on the threshold between materiality and immateriality, glass seemed to signify for Taut and Mies a repository of some profound expressive attributes more than simply being a construction material. Under the influence of Scheerbart, both saw glass as a contemplative substance that would give rise to psychic aspirations by engaging one with an imaginary insight. Read through under the guise of the French philosopher Gaston Bachelard's outlook, the two architects were after the poetry of materials that science destroyed and, correspondingly, engineers could not read. Specifically, the imaginary aura of glass made it exceptional among other materials and kept it outside the cognitive order. The potential of glass to act in the subjunctive mode of "as if" and to suspend material reality invited both architects to use it as a metaphor that opened the doors of the poetic reality. This is the common attribute of *Glashaus* and the Barcelona Pavilion that makes it possible to set up affinities between these two buildings.

To conclude, it is worth noting that, in writing *Glasarchitektur*, Scheerbart's aim was to create a *Traumkunst* (art of dream) but not a *Raumkunst* (art of space). His anticipation was that, in the twentieth century, it was neither reason nor faith that would break the eternal silence of infinite space, but the dream. With a similar optimist view, Bachelard suggested a return to the cosmology of dreams, covert in the primitive zone of "material reveries that precede contemplation."[39] As Empedocles stated, "it is through the earth that is in us that we know the earth, water through water, through our air the air divine, and through our fire devouring fire."[40]

For both Mies and Taut, buildings ought to signify more than a simple technical phenomenon (to which they have now been mostly reduced and quantitatively measured). Their search for pure expression did not intend to purify space of content and context, and thus also of meaning. They were interested in an iconoclasm that would permit one search for an invisible order behind the visible; and they replaced traditional stone with glass to explore two different ways of engaging with the world. Taut was interested in the vertical axis. He employed glass to elevate his building to a miracle-like inexhaustible force in order to cause a psychic excitement of airiness that would evoke the dreams of flight or fall. On the other hand, Mies was more attracted to the liquidity of glass. He used it to give a depth to the wall and make it unfathomable, like a mystery. The eyes, which dive into Mies's walls, were to look for pale and vague images of the self on a far horizon. Mies's portrait as the rational architect has been built upon his apathy for emotions. But, by using glass, he disclosed the material cause arising from the feelings.

Hidden behind the veil of reason, for the two architects, glass served as the perfect imaginative vehicle that allowed them to overturn the preeminence of reality over imagination. Deforming surrounding images, glass brought perception into the foreground and pushed consciousness into contemplation. With the two different metaphors it evoked for Taut and Mies, air and water, glass presents its polysemantic nature as an unconquerable poetic substance. "The less one knows, the more one names."[41]

BIBLIOGRAPHY

Anderson, Stanford, "*Sachlichkeit* and Modernity, or Realist Architecture," in H.F. Mallgrave (ed.), *Otto Wagner: Reflections on the Raiment of Modernity* (Santa Monica, CA: The Getty Center for the History of Art and the Humanities, 1993).

Bachelard, Gaston, *Water and Dreams: An Essay on the Imagination of Matter*, trans. E.R. Farrell (Dallas, TX: The Pegasus Foundation, 1983).

Behne, Adolf, "Bruno Taut," *Der Sturm* 4/198–9 (1914): pp. 182–3.

Blake, Peter, *The Master Builders: Le Corbusier, Mies Van Der Rohe, Frank Lloyd Wright* (New York, NY: W.W. Norton & Company, 1976).

Christofides, C.G., "Bachelard's Aesthetics," *The Journal of Aesthetics and Art Criticism* 20/3 (Spring 1962): pp. 263–71.

Cohen, Jean-Louis, *Mies Van Der Rohe* (London: Spon, 1996).

Dodds, George, *Building Desire: On the Barcelona Pavilion* (New York, NY: Routledge, 2005).

Evans, Robin, "Mies van der Rohe's Paradoxical Symmetries," in *Translations from Drawing to Building and Other Essays* (Cambridge, MA: The MIT Press, 1997).

Fischer, Theodor, "Was ich bauen moechte," *Kunstwart* 20/1 (October 1906): pp. 5–9.

Gottfried, Carl, "Hochhäuser," *Qualität* 3/5 (August 1922/March 1923): pp. 63–6.

Ikelaar, Leo, *Paul Scheerbart und Bruno Taut: Zur Geschichte einer Bekanntschaft: Scheerbarts Briefe der Jahre 1913–1914 an Gottfried Heinersdorff, Bruno Taut und Herwarth Walden* (Paderborn: Igel Verlag Wissenschaft, 1996).

Kaplan, E.K., "Gaston Bachelard's Philosophy of Imagination: An Introduction," *Philosophy and Phenomenological Research* 33/1 (September 1972): pp. 1–24.

Korn, Arthur, *Glas im bau und als Gebrauchsgegenstand* (Berlin: Ernst Pollak Verlag, 1929).

Lepik, Andres, "Mies and Photomontage, 1910–38," in Terence Riley and Barry Bergdoll (eds), *Mies in Berlin* (New York, NY: The Museum of Modern Art, 2001).

McGrath, R. and A.C. Frost, *Glass in Architecture and Decoration* (London: The Architectural Press, 1937).

Mendelson, David, *Le Verre et les Objets de Verre dans l'Univers Imaginaire de Marcel Proust* (Toulouse: Librairie José Corti, 1968).

Mertins, Detlef, "Architectures of Becoming: Mies van der Rohe and the Avant-Garde," in Terence Riley and Barry Bergdoll (eds), *Mies in Berlin* (New York, NY: The Museum of Modern Art, 2001).

Mies van der Rohe, Ludwig, n.t., *Frühlicht* 1/4 (1922): pp. 122–4.

Neumeyer, Fritz, *The Artless Word: Mies van der Rohe on the Building Art*, trans. Mark Jarzombek (Cambridge, MA: The MIT Press, 1991).

Partsch, Cornelius, "Paul Scheerbart and the Art of Science Fiction," *Science Fiction Studies* 29 (2002): pp. 202–20.

Scheerbart, Paul, "Licht und Luft," *Ver Sacrum* 1/7 (1898): p. 13.

Scheerbart, Paul, *Glasarchitektur* (Berlin: Verlag der Sturm, 1914).

Scheerbart, Paul, Bruno Taut and Dennis Sharp (ed.), *Glass Architecture by Paul Scheerbart and Alpine Architecture by Bruno Taut*, trans. James Palmes and Shirley Palmer (New York, NY: Praeger Publishers, 1972).

Taut, Bruno, "Eine Notwendigkeit," *Der Sturm* 4/196–7 (1914): pp. 174–5.

Taut, Bruno, *Die Stadtkrone* (Jena: E. Diedrichs, 1919).

Taut, Bruno, "Glaserzeugung und Glasbau," *Qualität; Wirtschaftliche Bildung und Qualitätsproduktion* 1/1–2 (April/May 1920): pp. 9–14.

Taut, Bruno, *Modern Architecture* (London: The Studio, Ltd., 1929).

Worringer, Wilhelm, *Formprobleme der Gotik* (Munich: R. Piper, 1912).

Worringer, Wilhelm and Herbert Read, *Form in Gothic* (New York, NY: Schocken Books, 1957).

NOTES

1   Paul Scheerbart, Bruno Taut and Dennis Sharp (ed.), *Glass Architecture by Paul Scheerbart and Alpine Architecture by Bruno Taut*, trans. James Palmes and Shirley Palmer (New York, NY, 1972), p. 14.

2   Ibid., p. 58.

3   One of these fiction writers was Marcel Proust, see David Mendelson, *Le Verre et les Objets de Verre dans l'Univers Imaginaire de Marcel Proust* (Toulouse, 1968).

4   Cornelius Partsch, "Paul Scheerbart and the Art of Science Fiction," *Science Fiction Studies* 29 (2002): p. 204.

5   Paul Scheerbart, "Licht und Luft," *Ver Sacrum* 1/7 (1898): p. 13.

6   Paul Scheerbart, *Glasarchitektur* (Berlin, 1914), p. 11.

7   The famous glass painter Gottfried Heinersdorff introduced Scheerbart to Taut in July 1913, when the architect was drafting the sketches of the glass pavilion. After

meeting Taut, Scheerbart sat to write *Glasarchitektur* as a programmatic account of the *Glashaus*, and they dedicated their respective works to each other. Leo Ikelaar, *Paul Scheerbart und Bruno Taut: Zur Geschichte einer Bekanntschaft: Scheerbarts Briefe der Jahre 1913–1914 an Gottfried Heinersdorff, Bruno Taut und Herwarth Walden* (Paderborn, 1996), pp. 87–144.

8   Bruno Taut, "Glaserzeugung und Glasbau," *Qualität; Wirtschaftliche Bildung und Qualitätsproduktion* 1/1–2 (April/May 1920): p. 9.

9   Ibid., p. 11.

10   Detlef Mertins, "Architectures of Becoming: Mies Van Der Rohe and the Avant-Garde," in Terence Riley and Barry Bergdoll (eds), *Mies in Berlin* (New York, NY, 2001), p. 114.

11   Fritz Neumeyer, *The Artless Word: Mies Van Der Rohe on the Building Art*, trans. Mark Jarzombek (Cambridge, MA: The MIT Press, 1991), p. 46.

12   Ludwig Mies van der Rohe, n.t., *Frühlicht* 1/4 (1922): pp. 122–4, translated in Neumeyer, *The Artless Word*, p. 240.

13   In the previous issue of *Frühlicht*, Behrens's, Mendelsohn's, Scharoun's and Taut's entries for the Friedrichstrasse Competition had been published.

14   Arthur Korn, *Glas im Bau und als Gebrauchsgegenstand* (Berlin, 1929), p. 7.

15   Taut wrote: "Human beings recover gradually their earth, and from this earth they make the carrier of their subtle feelings, the glass." Taut, "Glaserzeugung und Glasbau," p. 12.

16   Theodor Fischer, "Was ich bauen moechte," *Kunstwart* 20/1 (October 1906): p. 5.

17   Bruno Taut, "Eine Notwendigkeit," *Der Sturm* 4/196–7 (1914): p. 174.

18   Wilhelm Worringer, *Formprobleme der Gotik* (Munich, 1912).

19   Wilhelm Worringer and Herbert Read, *Form in Gothic* (New York, NY, 1957), p. 106.

20   Ibid., p. 106.

21   Ibid., p. 107.

22   Bruno Taut, *Die Stadtkrone* (Jena, 1919), p. 60.

23   Adolf Behne, "Bruno Taut," *Der Sturm* 4/198–9 (1914): p. 183.

24   As a noun that derives from the adjective *sachlich* and the noun *Sache*, *Sachlichkeit* has been translated in numerous ways: "objectivity," "thingness," "practicality," "straightforwardness," "functionalism," "realism" and "matter-of-factness." In the architectural discourse of the early twentieth century, *Sachlichkeit* referred to a norm used to measure the appropriateness of architecture to contemporary life conditions and was in general used to invoke "a straightforward attention to needs" missing in the world of daily life. Yet, practical reality and the range of needs that architecture addressed were phenomena open to diverse interpretations, which could be poles apart. Stanford Anderson, "*Sachlichkeit* and Modernity, or Realist Architecture," in H.F. Mallgrave (ed.), *Otto Wagner: Reflections on the Raiment of Modernity* (Santa Monica, CA, 1993), p. 340.

25   Walter Gropius was the organizer of the exhibition who declined Mies's Kröller-Müller project. Mertins, "Architectures of Becoming," p. 107.

26   Ibid., p. 114.

27   Bruno Taut, *Modern Architecture* (London, 1929), p. 71.

28   Mertins, "Architectures of Becoming," p. 117.

29   Mies, n.t., *Frühlicht*, p. 124.

30   Ibid.

31   Peter Blake, *The Master Builders: Le Corbusier, Mies Van Der Rohe, Frank Lloyd Wright* (New York, NY, 1976), p. 206.

32   Andres Lepik, "Mies and Photomontage, 1910–38," in Riley and Bergdoll (eds), *Mies in Berlin*, p. 325.

33   George Dodds, *Building Desire: On the Barcelona Pavilion* (New York, NY, 2005).

34   Ibid., p. 14.

35   Carl Gottfried, "Hochhäuser," *Qualität* 3/5 (August 1922 / March 1923): p. 63; Jean-Louis Cohen, *Mies Van Der Rohe* (London: Spon, 1996), p. 28.

36   R. McGrath and A.C. Frost, *Glass in Architecture and Decoration* (London, 1937), p. 370.

37   Robin Evans, "Mies van der Rohe's Paradoxical Symmetries," in *Translations from Drawing to Building and Other Essays* (Cambridge, MA, 1997), pp. 233–78.

38   Mertins, "Architectures of Becoming," p. 132.

39   Gaston Bachelard, *Water and Dreams: An Essay on the Imagination of Matter*, trans. E.R. Farrell (Dallas, TX, 1983).

40   C.G. Christofides, "Bachelard's Aesthetics," *The Journal of Aesthetics and Art Criticism* 20/3 (Spring 1962): p. 263.

41   Bachelard, *La Psychanalyse du Feu* (Paris, 1938), p. 70, quoted in E.K. Kaplan, "Gaston Bachelard's Philosophy of Imagination: An Introduction," *Philosophy and Phenomenological Research* 33/1 (September 1972): p. 4.

11

Found Spaces and Material Memory: Remarks on the Thickness of Time in Architecture

Jonathan Hale

Back in 2004 I was invited onto the competition jury to select an architect for what in 2009 became the new Nottingham Contemporary arts center. Resisting the understandable temptation to choose an international figure, the jury overlooked Zaha Hadid during the final stages in favor of what was then a little-known London practice called Caruso St John. A key factor in the final decision was Adam Caruso's evocative vision for the new gallery as a kind of "purpose-built found-space." His approach was based on the fact that many successful contemporary art spaces had been created within existing buildings. And not just any buildings: predominant among the precedents shown to the jury by Caruso were redundant industrial structures, marked with a distinctive individual history along with a family resemblance. His examples included: the Tate Modern in London, transformed by Herzog & De Meuron from the shell of the Bankside Power Station, and the Biennale galleries at the historic Arsenale in Venice, created from a complex of disused shipyards, armories and ropeworks. Other cases referred to included the Palais de Tokyo in Paris, a regeneration of a 1930s exposition building; MoMA's PS1 at Queens in New York based on a redundant neo-Romanesque public school; and the Museum for the Present, part of the Berlin National Gallery built inside a nineteenth-century railway station called the Hamburger Bahnhof. What all the examples had in common was an individual and powerful historical ambience: a sense of an abandoned shell of a former life somehow charged with future possibilities.

The peculiar preference for these qualities on the part of museum directors, curators and apparently many contemporary artists begs an obvious question: what is it about the reuse of redundant buildings that these people seem to find so compelling? One theme that often links these reconfigured structures is their connection with processes of production, whether directly as places of manufacture, or indirectly as nodal points in networks of infrastructure such as transport or power generation. Making and presenting art in these industrialized surroundings might therefore be an implicit response to Joseph Stalin's famous suggestion that creative writers and artists are "engineers of the human soul."[1]

More likely, the fascination with places of production may have emerged from a broader debate in museum and gallery design, one described by Tate director Nicholas Serota as the dilemma of experience versus interpretation. In his 1996 Walter Neurath lecture (also published as a book), Serota described a gradual shift in attitudes—from about the 1960s onwards—away from the curator-led model of the gallery as a place of viewing (under the top-down control of institutional conventions of display and interpretation) towards a more artist-led experiential environment where the gallery becomes almost an outpost of the studio. While discussing a number of site-specific pieces by the sculptors Carl Andre, Richard Serra and Joseph Beuys, Serota suggested that:

> These works also share a feature which has come to characterize the work of many artists of their generation: the sculptures were realized in the place of exhibition itself. The gallery or museum has become a studio, prompting a significant change in the conventional relationship between the artist, the work of art and the curator. No longer can the curator be seen solely as the dispassionate judge of quality, who visits the studio or private collection to select works and to assemble a body of material which will be presented to the public in a museum. Instead the curator is a collaborator, often engaging with the artist to accomplish the work.[2]

Alongside this burgeoning of site-specific and often installation-based production, there are also many cases where a deceased artist's own studio has been opened to the public as an exhibition space. In this situation the viewer is invited to contextualize the work in a very different way, often as a kind of witness to the artist's own habitual practices—their everyday routines and working methods as well as their sources of inspiration and innovation.

In both cases a notable aspect of the viewing experience is the presence of a temporal dimension, a sense of work having-been-made in a way that accommodates itself to the surroundings, or, in the case of the open studio, work-in-progress that might have been arrested during the process of production. Both of these dimensions of experience—the retrospective backward glance and the speculative forward projection—are also, intriguingly, key elements of a powerful conceptual model of the unfolding of time itself. One of the clearest formulations of this idea comes from within the phenomenological tradition, specifically in the work of Edmund Husserl on the "internal consciousness" of time.[3] Husserl suggested that our everyday notion of the present moment as an isolated instant in time is actually a misleading convention, as it is more precise and ultimately more productive to think of the lived present as a kind of composite of "retentions" and "protentions."[4] In other words there is a temporal thickness or layering that results from the gradual fading of the moment just passed alongside emerging anticipations of the future that is about to come. Rather than conscious memories or plans that would take much longer to bring "on line," these moments of past and future perception are an intrinsic part of the experience of the here and now—vital components of the sense of continuity or flow of experience. Perhaps the best illustration of this is in the experience of listening to music, where isolated sounds make no sense until they are strung together into a melody. It is only then that the rising and falling of notes begins to create a distinctive audible pattern, which along with a repeated cyclical rhythm gives a piece of music its unique identity—and partly why even a simple melody played in reverse is often completely unrecognizable.[5]

This sense of temporal depth and flow intrinsic to the "lived present" is also partly due to the time delays involved in the operation of the body's neurobiological and sensory-motor processing apparatus. The complexities of these interacting systems coupled with the endless flux of sensory information means that as the neuro-philosopher Daniel Dennett has forcefully explained there is never a moment in experience when we can say that "it all comes together." Dennett even went as far as to posit what he called a "multiple drafts model" of consciousness, in order to explain the presence of a number of partial and sometimes even contradictory versions of reality that seem to be circulating in our awareness at any given moment.[6]

Architectural corollaries to this notion of temporal "thickness" within the so-called present moment remain surprisingly under-explored. Notable exceptions include works addressing the persistence of a sense of history within the material remains of the past. An often overlooked twentieth-century example is *Pleasure of Ruins*, written by Rose Macaulay and first published in 1953.[7] More significantly there were of course a number of influential publications produced as a consequence of the 1960s reassessment of the typically modernist *tabula rasa* approach to the city, including Robert Venturi's *Complexity and Contradiction in Architecture* (1966), Aldo Rossi's *Architecture of the City* (1966), Kevin Lynch's *What Time is This Place?* (1972), and Colin Rowe and Fred Koetter's *Collage City*, which was conceived in 1973 but only published as a book in 1978. Perhaps as a consequence of positioning itself largely in opposition to the prevailing orthodoxy, much of this scholarship could be accused of preoccupation with nostalgia for the distant past. Hence it does not seem to take us very far in explaining the preference of contemporary artists for abandoned spaces, which, while exhibiting the relics of former uses and the historical traces of the lives of others, also seem to suggest that they are somehow pregnant with future possibilities.

FROM VITAL MATERIALITY TO MATERIAL TEMPORALITY

The approach taken here in the search to better understand what I would like to call the material imagination of time in architecture is to consider the current resurgence of interest in the notion of a "vital materiality." Across the human and social sciences over the past five to ten years there has been a discernible shift away from the previously dominant theoretical models: a backlash against the largely poststructuralist preoccupation with inter-textual analysis that has tended to dematerialize the world of things into a flux of "floating signifiers."[8] This earlier tendency could be compared with the effects of modern scientific methods that compulsively dissect and dissolve everyday objects into ever more miniscule sub-atomic elements. The American philosopher Graham Harman—a founder of the so-called "speculative realist" group—has described this as a simultaneous double movement of both "undermining" and "overmining."[9] Objects are either eroded from below by the atomizing effect of scientific analysis, or dissolved from above by a meta-narrative that describes them as nothing more than the result of their relations—positing them as cultural constructions or effects of social discourse. The consequence of both approaches is to obscure the significance of everyday experience and its encounter with the sheer brute reality of material things.

Harman's so-called "object oriented ontology" tries to imagine a world where objects encounter each other away from the gaze of human beings, or at least where human perception has no special privilege over other forms of material interaction. In fact, according to Harman's description of the "as-structure" of experience,[10] we humans are constantly caricaturing the objects we encounter by taking them *as* something more or less useful to our goals and thereby overlooking many of their other less immediately salient qualities. Likewise even inanimate objects could also be said to grasp each other only partially, according to the particular qualities with which they are able to interact. Thus a fire could be said to encounter only the combustibility of a piece of cotton, or a rock to interact only with the breakability of a pane of glass. This logic ultimately generates a distinctively flat ontology, one that puts human beings on a similar level to material objects.

This position echoes very strongly the understanding of technical systems advocated by Bruno Latour, the French philosopher and sociologist of technology with whose ideas Harman has frequently engaged.[11] Latour has suggested that the difficulty of analyzing technological "objects" is due to their mixing of human and non-human "actors," and the ways in which they throw together complex networks of interacting forces that defy conventional categorization. In much of Latour's writing these confusing hybrids are painstakingly teased apart through detailed empirical case studies of particular technical phenomena. Notable examples include his analysis of Louis Pasteur's discovery of microbes,[12] and the history of an abandoned French mass-transit project.[13] But on a broader philosophical level he has also tried to explain the conceptual difficulty of dealing with technological systems, which necessitates working across the kind of binary categories that are embedded in our everyday patterns of thought. The problem he explains lies at the heart of the "modern project," which attempts to make a series of conceptual distinctions that follow an either–or pattern. An object is defined as either, natural or cultural, mental or physical, living or non-living, animal or vegetable etc., none of which leaves room for the kind of messy and tangled hybrids that technology constantly throws up—such as steam trains, mobile phones, armies or insurance companies. More pressingly in light of current developments in biotechnology such as gene therapy and prosthetics, the blurred boundary between the human and the technological is also impossible to grasp in binary terms.

This notion of a posthuman condition (broadly, the latest phase in the continuing erosion of the traditional humanist subject-centered world view) is now being explored by a number of current thinkers in terms of its broader social and political implications. Latour, in typically candid fashion, has provided a vivid and down-to-earth example of the blurring of subject and object in his patient (and often humorous) analysis of a humble door-closer mechanism. In one of his few essays addressing a specifically architectural theme he shows how even the functioning of a simple laborsaving device like a hydraulic door-closer can involve a complex interplay of human and non-human agency.[14] Alongside the deliberate devolution of a human ability (closing a door) into the operation of a mechanism, there are also many ways in which the brute materiality of the object can still manage to reassert itself. Whether by shutting too quickly and trapping people's fingers or being too strong for a child to open, the device soon takes on an individual personality, much as if a living concierge were actually opening the door according to the mood of the moment. In all these examples Latour's approach demonstrates the active contributions made by physical materials, questioning the conventional view of matter as dead weight or inertia that once overcome can then be forgotten.

Returning to the theme of the posthuman, but in relation to Latour's idea of material agency, an important new wave of scholarship has begun to emerge in this area written from a broadly feminist perspective. This trend picks up on an earlier preoccupation in the social sciences with the cultural and political status of the human body, a movement initially inspired by Michel Foucault and reaching a peak of interest in the 1980s. This earlier "anti-essentialist" approach attempted to throw into question binary categories such as gender classification, revealing the influence of cultural forces in the acquisition of behavioral norms. This kind of analysis tended to obscure the material capacities of human embodiment as a potential source of individual agency, instead rendering the body as a basically passive victim of the larger forces of social inscription. As Foucault himself has claimed: "power relations have an immediate hold on [the body]; they invest it, mark it, train it, torture it, force it to carry out tasks."[15] But, if this were the case, it would place this notoriously shadowy notion of power somehow prior to embodiment, when in reality bodies must already be implicated in the very processes by which power relations are produced.

Rather than focusing simply on the materiality of the body-as-such this new writing takes up the broader theme of material embodiment in general, but as with the earlier work mentioned above it is driven by an interest in its social and political implications. An important recent collection of this writing is edited by Diana Coole and Samantha Frost, called *New Materialisms: Ontology, Agency, and Politics*.[16] Much of this work contains references to key figures from the poststructuralist canon mentioned already while also making connections back to thinkers from the earlier phenomenological tradition. Drawing on the work of Gilles Deleuze and Michel Foucault, as well as philosophers of technology like Donna Haraway, the political theorist Rosi Braidotti draws attention to the problematic boundary conditions that now exist in relation to the definition of the posthuman. These include the edges of life (beginnings and ends); the edges of the species (cloning and gene-splicing); the edges of the self (prosthetic extensions or medical invasions); and even what this blurring of boundaries might mean for the future definition of the humanities itself as an academic discipline.

In contrast, the writing of Jane Bennett that is also included in the above collection returns to more historical sources for clues to what she calls a "vital materialism."[17] Interested in the difference between the passive and active aspects of nature, she takes from Baruch Spinoza the distinction between *natura naturata* and *natura naturans*—the former refers to nature as a set of fixed and ordered forms while the latter suggests the power of nature as a creative generator of new possibilities.[18] She relates this idea to the so-called élan *vital* or creative force described in the later work of Henri Bergson although she attempts to avoid the accusation of mysticism often leveled in his direction. Without admitting to a full-blown vitalism she pursues the idea of a "material propensity," a kind of natural tendency inherent in a material that suggests how it might best be employed.[19] Bennett goes on to compare this with the notion of *Shi* found in ancient Chinese thought, as recently described by the French historian François Jullien in a book called *The Propensity of Things*:

> *Shi is the style, energy, propensity, trajectory, or élan inherent to a specific arrangement of things. Originally a word used in military strategy, shi emerged in the description of a good general who must be able to read and then ride the shi of a configuration of moods, winds, historical trends, and armaments: shi names the dynamic force emanating from a spatio-temporal configuration rather than from any particular element within it.*[20]

As with Latour's work mentioned earlier, Bennett takes this "dynamic force" as an incipient form of agency, one that emerges often unpredictably in the operation of a complex technical system. One of her key examples is the North American electrical power network, illustrated most dramatically in the famous cascade of failures that resulted in the widespread blackouts of August 2003.[21] At the same time Bennett also traces some of the sources of the deeper cultural significance of particular materials and processes, in a way strongly reminiscent of Gaston Bachelard's pioneering work from the 1940s on the material imagination of the four elements.

This idea of emergent agency is taken to an even more radical conclusion by Diana Coole, one of the editors of *New Materialisms*, who returns to the later writings of Maurice Merleau-Ponty—specifically his recently published lectures on the concept of nature given at the Collège de France in the 1950s. Referring to the biologist Jakob von Uexküll's idea of the *Umwelt* (broadly, the particular characteristics of an environment with which a given organism is able to interact), Merleau-Ponty described how this idea "rejected the model of the organism as a physical machine animated by consciousness or by some vital spark, describing instead an emergent future-oriented but open organisation that is immanent to the organism."[22] Merleau-Ponty's own suggestion that this "behavioral activity oriented toward an *Umwelt* begins well before the invention of consciousness,"[23] could also now be read in light of the work of Graham Harman referred to above, whose ideas imply the presence of a minimal form of subjectivity in the capacity of material objects to take account of each other's qualities.

Another contributor to *New Materialisms* who also draws on the work of Merleau-Ponty is the political theorist William Connolly. In an essay entitled "The Materiality of Experience" he explores the presence of a time lag in the process of perception, as identified by the neuropsychologist Benjamin Libet in a series of experiments carried out in the 1970s.[24] Libet's findings suggested a roughly half-second delay between the arrival of incoming sensory stimulation and the formation of a perceivable experiential image, whether visual, auditory, tactile and so on.[25] Connolly takes this as support for Merleau-Ponty's idea of embodied perception as an ongoing sequence of interactions with the world, where the materiality of the body itself makes a key contribution to the characteristic quality and texture of experience.[26] In the time taken for a perceptual image to form, information shuttles back and forth across the body–world boundary, and thus it becomes impossible to say categorically which qualities belong to which. Merleau-Ponty's paradigmatic example of this is the body touching itself, such as when the right hand grasps the left hand and a perceptual ambiguity immediately occurs.[27] He claims this confusion results from the reversibility between the "seer" and the "seen," that is, the idea that in order to experience a world at all we must share in its materiality through the thickness of our own embodiment. In other words, the qualities of the things in the world that we experience through our bodies are, by definition, just what-it-feels-like to experience them.

THE THICKNESS OF EMBODIMENT IN ART AND ARCHITECTURE

Returning to the question I began with, as to why artists and museum directors seem so peculiarly drawn to found spaces, I would like to propose that some of this attraction could be explained by reference to other forms of architectural recycling. There is now a growing body of historical scholarship on the creative use of *spolia*, referring to the practice of removing and reusing constructional elements such as columns, doorways or friezes. While there is evidence of this happening even in Classical times it became more widespread during the Middle Ages, where it seems to take on a similar significance to the Christian reappropriation of Pagan festivals—a move that according to one recent scholar "encompasses both rejection and continuation."[28] Dalibor Vesely has likewise written extensively on the broader significance of the historical fragment, relating it to the origins of the museum:

> The restorative or symbolic meaning of the fragment can be discerned already
> in the spoglia (spoils) so frequently used in the Middle Ages—equally in the
> collections of curiosities of the late Renaissance, or in the cult and poetics of ruins,
> which reached a peak in the eighteenth century.[29]

In an attempt to assimilate some of these ideas within a more general theory of reappropriated spaces, Fred Scott in his recent book *On Altering Architecture*, makes a number of important observations. While the reuse of *spolia* involves a dynamic process of juxtaposing fragments brought from different contexts, Scott is more interested in the effects produced by preserving and reusing elements in situ. Likening the process of "intervention design" to the techniques of Synthetic Cubism and Surrealist collage, he also references the sculptural work of Gordon Matta-Clark and his full-scale dissections of domestic spaces.[30] Here Scott also emphasizes the experiential dimension of this kind of opening up of new spatial connections, liberating previously unavailable viewpoints and offering new possibilities of movement: "Such an imposition of a new spatial and circulatory hierarchy will allow the same privileges previously available only to thieves or ghosts, that is a novel view of the original hierarchy, which will now be a relic and memorial of a previous occupation."[31]

Another of Scott's observations that hints at the temporal dimension I have been trying to develop is the comparison of these spatial transformations with the process of transcription and translation. While in broad terms this may indeed involve "the carrying over of the host building from one age to another,"[32] I would like to suggest that the powerful sense of time perceptible in the best examples of creative reuse might actually result from the narrative gaps that appear between the traces of old and new uses. In the most interesting examples referred to earlier where there is a shift from one function to another, a kind of disjunction or contradiction appears between the visual and functional form. On the one hand, there are those "physical propensities" that emerge when a material is being worked, such as the resistance of a block of stone to being chiseled into a building component. These are what give a specific material its particular tectonic qualities. On the other hand, there are spatial and functional propensities that also suggest how a building might be used. Reappropriation of spaces therefore involves a layering of successive phases of transformation, where the building is less a tectonic object and more a series of tectonic events. Each one of these events involves unpredictable displacements that can leave their mark within the building fabric, and a typical sequence would include: transformations from raw material to building component; from building component to finished building; from finished building to inhabited space; from inhabited space to redundant building; from redundant building to re-programmed building; and finally from re-programmed building to re-occupied space. The presence of gaps or displacements between each of these successive layers may be what allows for new possibilities of signification and meaning to emerge, although it is not yet clear how this process of architectural codification takes place.

One clue to the mechanism by which new significations can be captured comes from the work of Merleau-Ponty on the process of innovation in language. To explain this idea he drew a distinction between what he called "spoken" and "speaking" speech, where the former refers to the more familiar and well-worn patterns of conventional everyday language. Speaking speech on the other hand describes the more challenging and rarefied patterns of poetic expression, where we often experience a sense of estrangement from conventional meanings as if the writer is deliberately playing with the possibilities of the language. Sometimes this can literally involve reconfiguring and distorting existing forms of expression as a way of capturing, and to some extent actually producing new levels of meaning, as Merleau-Ponty himself suggests: "It is just this process of 'coherent deformation' of available significations which arranges them in a new sense and takes not only the hearers but the speaking subject as well through a decisive step."[33] Or in other words, in taking up and making do with the limited resources that language offers (a common currency invented by others to satisfy other—albeit related— purposes), we are constantly searching for ways to inflect our speech to suit the unique and fleeting demands of the moment. While language operates on a "deficit and surplus" model, expression will always fall short of the speaker's intention, but there is at the same time an unexpected excess in the historical associations that words carry with them: the sedimented layers of meaning accumulated over years of use will ensure that the speaker also says more than was actually intended. And thus, in thinking out loud, we are often able to clarify even a half-formed feeling. As Merleau-Ponty pointed out: "my spoken words surprise me myself and teach me my thought."[34] It is this sense of historical excess embedded within the thickness of the medium that I would like to suggest is a key aspect of the richness of potential that found spaces seem able to offer.

In a final effort to better understand these spaces, and in particular why they seem so peculiarly pregnant with possibilities of appropriation, it is useful to see how Merleau-Ponty's ideas might apply to the materiality of other media. A good example is the work of the new media and performance theorist Carrie Noland and her recent book entitled *Agency and Embodiment*.[35] In this text she makes direct reference to Merleau-Ponty's work on perception in her discussion of the multimedia artist Bill Viola, specifically his time-lapse video piece from 2000 entitled *The Quintet of the Astonished*.[36] In this work five actors are filmed performing a sequence of facial expressions, communicating the canonical emotional states of fear, anger, pain, sorrow and joy. By shooting the video at up to 384 frames per second instead of the typical 24, one minute of live action is extended to 16 minutes of viewing time, allowing previously unnoticeable movements to become visible. Viola suggests that this blurs the normally obvious distinctions between one emotional expression and another, while opening up for the viewer's inspection the previously unseen transitions between them. In this way a whole set of ambiguous new expressions becomes available to be assigned to new meanings.

The lesson to be taken from this, as with all the examples discussed, is that the embodied physicality of materials is what puts them beyond our complete control. And likewise it is our own bodily materiality that puts us beyond the reach of power. Whatever our language of expression, whether brain, body or building (new or old), it is the thickness of the medium itself that holds the potential of temporality—the promise of a productive registering of both posterity and possibility.

BIBLIOGRAPHY

Bennett, Jane, *Vibrant Matter: A Political Ecology of Things* (Durham, NC: Duke University Press, 2010).

Coole, Diana and Samantha Frost (eds), *New Materialisms: Ontology, Agency, and Politics* (Durham, NC: Duke University Press, 2010).

Dennett, Daniel, *Consciousness Explained* (1991; London: Penguin, 1993).

Derrida, Jacques, *Writing and Difference*. Translated by Alan Bass (London: Routledge & Kegan Paul, 1978).

Fabricius Hansen, Maria, *The Eloquence of Appropriation: Prolegomena to an Understanding of Spolia in Architecture* (Rome: L'Erma di Bretschneider, 2003).

Foucault, Michel, *Discipline and Punish: The Birth of the Prison*. Translated by Alan Sheridan (New York, NY: Vintage, 1995).

Harman, Graham, *The Quadruple Object* (Winchester: Zero Books, 2011).

Harman, Graham, *Towards Speculative Realism: Essays and Lectures* (Winchester: Zero Books, 2010).

Husserl, Edmund, *On the Phenomenology of the Consciousness of Internal Time (1893–1917)*. Translated by J. Brough (Dordrecht: Kluwer, 1991).

Latour, Bruno, *Aramis, Or the Love of Technology*. Translated by Catherine Porter (Cambridge, MA: Harvard University Press, 1996).

Latour, Bruno, "Mixing Humans and Nonhumans Together: The Sociology of a Door-closer." *Social Problems* 35/1 (June 1988): pp. 298–310.

Latour, Bruno. *The Pasteurization of France*. Translated by Alan Sheridan and John Law (Cambridge, MA: Harvard University Press, 1988).

Macaulay, Rose, *Pleasure of Ruins* (1953; London: Thames and Hudson, 1984).

Massumi, Brian, *Parables for the Virtual: Movement, Affect, Sensation* (Durham, NC: Duke University Press, 2002).

Merleau-Ponty, Maurice, *Nature: Course Notes from the College de France*. Translated by Robert Vallier (Evanston, IL: Northwestern University Press, 2003).

Merleau-Ponty, Maurice, "On the Phenomenology of Language." In *Signs*. Translated by Richard C. McCleary (Evanston, IL: Northwestern University Press, 1964), pp. 84–97.

Merleau-Ponty, Maurice. *Phenomenology of Perception*. Translated by Donald A. Landes (Abingdon: Routledge, 2012).

Morris, David, *The Sense of Space* (Albany, NY: SUNY Press, 2004).

Noland, Carrie, *Agency and Embodiment: Performing Gestures/Producing Culture* (Cambridge, MA: Harvard University Press, 2009).

Scott, Fred, *On Altering Architecture* (Abingdon: Routledge, 2008).

Serota, Nicholas, *Experience or Interpretation: The Dilemma of Museums of Modern Art* (1996; London: Thames & Hudson, 2000).

Veseley, Dalibor, *Architecture in the Age of Divided Representation: The Question of Creativity in the Shadow of Production* (Cambridge, MA: MIT Press, 2004).

Westerman, Frank, *Engineers of the Soul: In the Footsteps of Stalin's Writers* (New York, NY: Vintage, 2011).

NOTES

1   Joseph Stalin, "Speech at the Home of Maxim Gorky," 26 October 1932. Quoted in Frank Westerman, *Engineers of the Soul: In the Footsteps of Stalin's Writers* (New York, 2011), p. 34.

2   Nicholas Serota, *Experience or Interpretation: The Dilemma of Museums of Modern Art* (1996; London, 2000), p. 36.

3   Edmund Husserl, *On the Phenomenology of the Consciousness of Internal Time (1893–1917)*. Translated by J. Brough (Dordrecht, 1991), p. 11.

4   For a useful analysis of Husserl's account see Maurice Merleau-Ponty, *Phenomenology of Perception*. Translated by Donald A. Landes (Abingdon, 2012), pp. 439–42.

5   A principle that the British composer Mike Oldfield famously exploited to great effect in basing his multi-platinum selling album *Tubular Bells* on the opening sequence of Bach's Toccata and Fugue in D minor played backwards.

6   Daniel Dennett, *Consciousness Explained* (1991; London, 1993), pp. 101–38.

7   Rose Macaulay, *Pleasure of Ruins* (1953; London, 1984).

8   Jacques Derrida, *Writing and Difference*. Translated by A. Bass (London, 1978), p. 25.

9   Graham Harman, *The Quadruple Object* (Winchester, 2011), pp. 7–19.

10 Graham Harman, *Towards Speculative Realism: Essays and Lectures* (Winchester, 2010), p. 36.

11 See, for example, the essay "Bruno Latour: King of Networks," in Harman, *Towards Speculative Realism*, pp. 67–92.

12 Bruno Latour, *The Pasteurization of France*. Translated by A. Sheridan and J. Law (Cambridge, MA, 1988).

13 Bruno Latour, *Aramis, Or the Love of Technology*. Translated by C. Porter (Cambridge, MA, 1996).

14 Bruno Latour, "Mixing Humans and Nonhumans Together: The Sociology of a Door-closer," *Social Problems* 35/1 (June 1988): pp. 298–310.

15 Michel Foucault, *Discipline and Punish: The Birth of the Prison*. Translated by Alan Sheridan (New York, NY, 1995), p. 25.

16 Diana Coole and Samantha Frost (eds), *New Materialisms: Ontology, Agency, and Politics* (Durham, NC, 2010).

17 Jane Bennett, *Vibrant Matter: A Political Ecology of Things* (Durham, NC, 2010), p. vii.

18 Ibid., p. 117.

19 Ibid., pp. 60–61.

20 Ibid., p. 35.

21 Ibid., pp. 24–8.

22 Ibid., p. 103.

23 Maurice Merleau-Ponty, *Nature: Course Notes from the College de France*. Translated by R. Vallier (Evanston, IL, 2003), p. 167.

24 William E. Connolly, "Materialities of Experience," in Coole and Frost, *New Materialisms*, pp. 178–200.

25 For more detail on the experiments, alongside discussion of their cultural implications, see: Brian Massumi, *Parables for the Virtual: Movement, Affect, Sensation* (Durham, NC, 2002), pp. 28–31.

26 In reference to Merleau-Ponty the philosopher David Morris has recently described this process as "the crossing of body and world." David Morris, *The Sense of Space* (Albany, NY, 2004), pp. 4–5.

27 Merleau-Ponty, *Phenomenology of Perception*, pp. 94–5.

28 Maria Fabricius Hansen, *The Eloquence of Appropriation: Prolegomena to an Understanding of Spolia in Architecture* (Rome, 2003), p. 260.

29 Dalibor Veseley, *Architecture in the Age of Divided Representation: The Question of Creativity in the Shadow of Production* (Cambridge, MA, 2004), p. 322.

30 Fred Scott, *On Altering Architecture* (Abingdon, 2008), pp. 127–33.

31 Ibid., p. 154.

32   Ibid., p. 79.

33   Maurice Merleau-Ponty, "On the Phenomenology of Language," in *Signs*. Translated by R.C. McCleary (Evanston, IL, 1964), p. 91.

34   Ibid., p. 88.

35   Carrie Noland, *Agency and Embodiment: Performing Gestures/Producing Culture* (Cambridge, MA, 2009).

36   Ibid., pp. 66–72.

12

The Return of the Ruin: Modernism, History and the Material Imagination

Jonathan Hill

"Bomb damage is itself picturesque," said Kenneth Clark—director of the National Gallery and chairman of the War Artists Advisory Committee—following the aerial bombardment of London during the Blitz.[1] Clark's stoic embrace of ruination was in line with a burgeoning romanticism in 1940s Britain that celebrated national identity as a way of bolstering the nation against military aggression. The tenets of this romanticism soon found support among other figures of Britain's intelligentsia, with Clark joined by T.S. Eliot and John Maynard Keynes in writing a letter to *The Times* in 1944 in which they stated that a ruined church would be an evocative monument to wartime sacrifices.[2] Their letter was reprinted in a subsequent publication, *Bombed Churches as War Memorials* (1945), in which the landscape architect Brenda Colvin complemented this recognition of the cultural value of the damaged ruin with a corresponding call for an enveloping and unkempt nature. Her landscape proposal for Christopher Wren's Christ Church, Newgate Street, would "emphasize the passing seasons" in relation to the "charred and battered" church and "the crisp polished facades of the surrounding buildings," reintroducing "the self-sown flowers" that had flourished during the sustained German bombing raids of 1940 and 1941.[3] Returning to this theme in the second edition of *Land and Landscape*, 1947, she wrote: "With a little imagination one might visualize a London left to nature's healing hand after all mankind was doomed, and see, in the mind's eye, a lost and broken city hidden under a great forest of sycamore."[4]

The return of the ruin in mid-twentieth-century Britain was a reassertion of a familiar, evocative theme. The British concern for ruination came to fruition in the eighteenth century due to empiricism's attention to subjective experience, the heightened historical awareness in the Enlightenment's concern for origins and archaeology, and the value given to imagination, time and metaphor in the picturesque and romanticism. A massively monumental ruin was assumed to exemplify the majesty and emotive power of architecture more eloquently even than a complete building. Diminishing objects physically, ruins were understood to expand their metaphorical potential: "for imperfection and obscurity are their properties; and to carry the imagination to something greater than is seen, their effect," wrote Thomas Whatley in 1770.[5] Rather than only associate the immaterial with timeless, ideal geometries, the eighteenth century increasingly conceived the immaterial as temporal and subject to experience, not only in the actual absence of matter, but also in the perceived absence of matter seen through mists and storms, establishing a dialogue between the immaterial and material that associated self-understanding with the experience of objects subject to nature and weather. The ruin represented growth as well as decay, and potential as well as loss. To contemplate the past was also to imagine the future, and ruination was often a precursor to change. Evoking life and death in a single object, the ruin of a building was linked to the ruin of a person or a place, as well as their potential for renewal. Eighteenth-century Britain became so associated with its landscape that the ruin provided a means to negotiate between culture and nature and was synonymous with the fluctuating fate of the nation, establishing an emotive tradition that acquired new resonance in the mid-twentieth century.

Proposing a vocabulary for post-war reforestation in *Trees for Town and Country*, 1947, Colvin notes that just 23 out of 60 trees originated in Britain and invokes an allegory of liberalism: "Although introduced to Britain by human agency, the Spanish chestnut grows well on light soils and suits our landscape. It has become so well integrated that the eye accepts it as a native tree."[6] Nikolaus Pevsner, the most sustained advocate of the twentieth-century picturesque, and himself something of a non-native transplant, offered a human equivalent to Colvin's sylvan allegory: "England has indeed profited just as much from the un-Englishness of the immigrants as they have profited from the Englishing they underwent."[7] The diverse origins of the picturesque and its openness to new influences were important in the eighteenth century and again two centuries later, when at the height of the war Pevsner recalled the traditional two-way cultural dialogue between England and continental Europe, describing the picturesque as England's principal contribution to European architecture.[8] For Pevsner, the picturesque was "tied up with English outdoor life and ultimately even the general British philosophy of liberalism and liberty."[9]

The themes of national remembrance and renewal that Clark, Colvin and Pevsner emphasized in the 1940s later combined with the optimism of the postwar welfare state to evoke both egalitarian monumentality and thoughts of past and future ruination, recalling the Blitz and acknowledging the new threat of the Cold War. The idea of a British welfare state was established in 1942 when an inter-departmental committee chaired by Sir William Beveridge presented its report to the wartime collation government. The Labour party's landslide victory in 1945 brought its conclusions to fruition, and the welfare state as realized aimed to extend access to good schools, universities and hospitals to the whole population, but did not intend a fundamental transformation of capitalism or attempt to address financial inequalities between rich and poor.

In 1961 the Conservative government set up a Committee on Higher Education chaired by Professor Lord (Lionel) Robbins, which after two years deliberation recommended the creation of a number of new universities. In fact, by the time of their report, seven new universities were already in development under the remit of the Universities Grants Committee (UGC). Emphasizing the coexistence of teaching and research in a residential community, these "plateglass" universities were located in smaller provincial towns and cities of a comparable size and character to Oxford and Cambridge rather than major cities, the location of civic and redbrick universities, which had larger student populations and a more dispersed academic community. The first in Britain to be fully controlled by the national government and synonymous with the welfare state, the new universities aimed to extend access to all people with "the qualifications and the willingness to pursue higher education," as the Robbins Report concluded.[10]

As support for new universities grew, the initial working title for one of the seven—University of Norwich—was discarded because the local Promotion Committee believed that a university with regional backing was more likely, and the UGC approved the University of East Anglia (UEA) in 1960. Of all the cities that acquired a new university, Norwich was the most isolated, literally as well as socially. In an obituary for Bert Hazell, who had been the Labour MP for North Norfolk, Edward Coke, seventh Earl of Leicester and owner of the local Holkham estate, remarked: "I respected Hazell. He was the MP who came into the one constituency in England where, in 1964, it was so feudal that it had to be explained to the electors that the ballot was secret."[11]

The focus of the UEA campus, Earlham Hall, was a large house with a sixteenth-century core and later additions. Once owned by a descendant of the philosopher Francis Bacon—the father-figure of empiricism, the principal British contribution to Enlightenment theory—the Hall was later leased to the Gurneys, an influential Quaker family who in the 1790s created a picturesque park along the River Yare, which meanders to the west and south of the house. In a decade known for cultural and social experimentation, a picturesque setting in eighteenth-century parkland was apposite for a new university precisely because of its association with British liberalism. In Norfolk, such a site had special resonance because the county is known for some of England's grandest eighteenth-century estates, including Holkham and Houghton, home of the first British Prime Minister, Sir Robert Walpole. UEA's first Vice-Chancellor, Frank Thistlethwaite, remarked of Earlham: "There emanated from its associations with the Bacon and Gurney families and the Norwich circle a *genius loci* which could not fail to imbue students with a sense of their cultural inheritance."[12] The long association of the picturesque with liberty and its reassertion in the 1940s and 1950s as a totem of national identity, which Pevsner associated with a local, empirical and environmentally aware modernism, ensured the appropriate combination of tradition and innovation.[13] While eighteenth-century England advocated liberalism, only a small proportion of the population was allowed a university education, the right to vote, and access to a picturesque estate. In contrast, the welfare state aimed to open these rights and pleasures to the whole population.

Soon after his appointment in 1962 as architect of the new university, Denys Lasdun toured Norfolk, visiting Holkham, Houghton and Humphry Repton's park at Sheringham. With regard to Earlham, he emphasized his intention "not to wreck for all time the most wonderful landscape in which we find ourselves … it's a very, very beautiful place."[14] According to Repton: "The spot from whence the view is taken, is in a fixed state to the painter; but the gardener surveys his scenery while in motion;

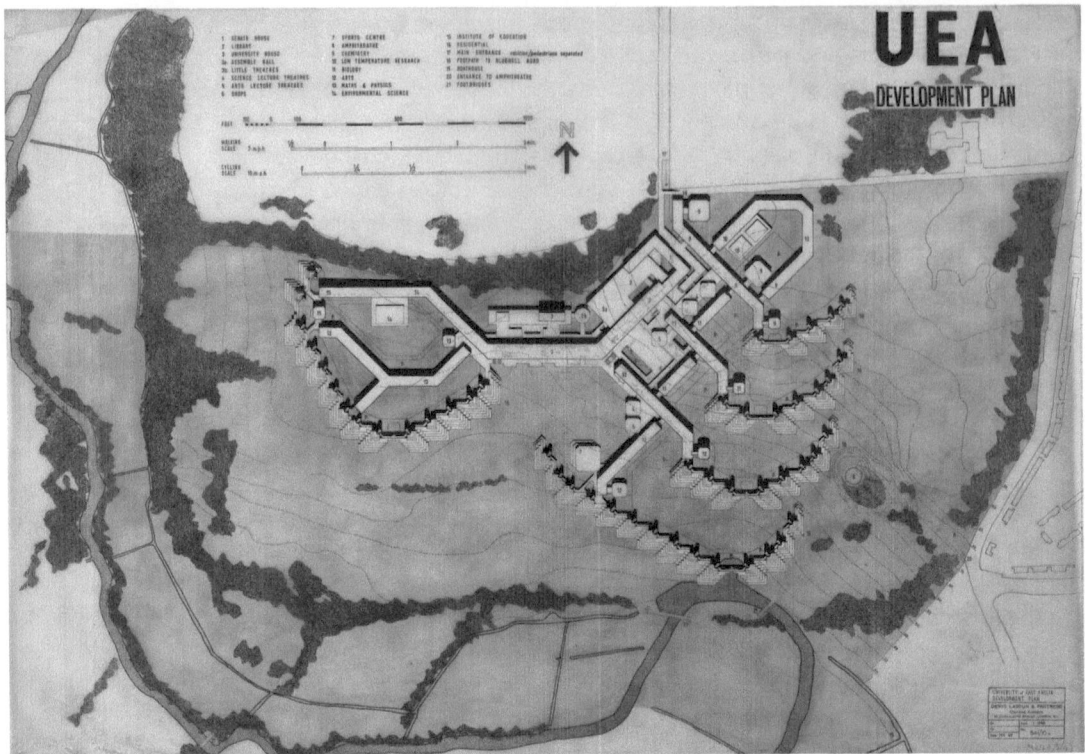

12.1   Denys
Lasdun, UEA
Development Plan,
1962.

and … sees objects in different situations."[15] Recalling this practice, Lasdun walked the site "in all seasons, in mist, snow, wind and sun."[16] In a key design decision with picturesque connotations, he decided that the various architectural elements "were to be disposed on this site with loving care for the configuration and contours of the landscape, its prospect and aspect," and concluded: "If Repton had been asked to do this university on this landscape, he would have said, 'Keep it that way.'"[17] Repton had no involvement with Earlham but he was an appropriate precedent, having lived and worked in Norfolk for over 20 years and designed Sheringham as well as a garden for Bartlett Gurney at Northrepps, north of Norwich, in the early 1790s. Lasdun's reference to Repton indicated not only the potential he saw in the site but also his empathy for East Anglia and England. In an early press conference he advocated a regional as well as national architecture: "There will be no question of trying to create a university for Norwich at long distance … The greater modern movements fired my youth but now I am past my youth and I think for myself … I am very much concerned with rather special values, solely, or mostly, applying to England."[18]

Remarking that "I became interested in designing buildings which responded almost ecologically to unique and specific situations," Lasdun recommended Colvin as UEA's landscape architect.[19] Advocating a holistic appreciation of nature, she recognized a "contact or communication between the individual mind and the universal subconscious, through means (which I believe to be) natural and physical."[20] Hal Moggridge, who became her business partner in 1969 and collaborated on UEA, characterizes Colvin as "a romantic, as am I."[21] Confirming Lasdun's ecological intentions, Colvin maintained the site's rich variety of natural habitats, including hedgerows, marshes, meadows, riverbanks and woods.[22] As far as possible, she wanted the landscape to be "a self-conserving system," remarks Moggridge.[23] The rising ground was made to gently dip close to the base of Lasdun's residential ziggurats so that they appear to rise directly from the land like the rocky outcrops that the architect intended (Figure 12.1). Employing coastal, landscape and urban metaphors, Lasdun remarked that the "architectural hills and valleys" of an "academic city" are an "outcrop of stone on the side of a hill leading down to a river" and "landlocked harbour."[24] Nearer to the university buildings, Colvin recommended that fine grass would be closely mown, while further down the slope it would be of a rougher texture and left long and only "scythed occasionally," contrasting a cultivated lawn to a wild meadow.[25]

Already in 1957 Lasdun had remarked that his intention was to create "high intensity, monumental, poetic buildings."[26] In the most frequently published photograph of UEA, the ziggurats' craggy silhouette is seen across the marshy meadow and against the misty evening glow of a vast Norfolk sky scattered with high drifting clouds (Figure 12.2). To the left, the sparse foliage of a large tree frames the view, very much in the manner that the seventeenth-century painter Claude would frame his own subjects through left-of-center overhanging leaves

12.2   Denys Lasdun, UEA, 1968, View from the River Yare.

and branches, inspiring both eighteenth-century and twentieth-century advocates of the picturesque.[27]

Whether in situ for the teaching spine or precast for the residential ziggurats, UEA's predominant material is concrete. Seen from the north, the student residences offer towering blank walls and a castellated profile. Associations with bastions, bunkers and bomb shelters are unavoidable. To his son, Lasdun remarked: "You know James, there's something aphrodisiacal about the smell of wet concrete."[28] But he explained his choice in relation to its context: "Of all the suitable materials available today concrete in its natural grey state appears to enhance the colours of the landscape to the greatest advantage."[29] Students and academics in the 1960s remember that the fresh, pale concrete offered a strong contrast with the lush, green grass.[30] But Susan Lasdun remarks that her husband was "very romantic about rain-streaked concrete."[31] Marked with algae, lichen and moss, UEA appeared archaic by the end of the twentieth century, suggesting wartime destruction as well as picturesque and ancient ruins. Fecund in a damp climate, dappled fungal blotches offered seasonal variation, merging with the blacks and grays of winter trees and contrasting with nature's vibrant summer colors. Algae, lichen and moss were consistent with Lasdun's conception of the ziggurats as rocky outcrops tucked into the slope above a moist, marshy river, and he enjoyed the romantic appeal of UEA's later appearance. Of the National Theatre, London, 1976, he remarked, "I have always wanted to see the exterior with something growing on it—Virginia creeper would be ideal, changing color with the seasons," adding, "It will weather, it will streak, it will become part of nature. It will probably get lichen from the river, there will be trees around it."[32] Of UEA, he concluded, "As bits get chipped off and bits grow around it, I think it will become part of landscape … On a wet day it may look drab and forbidding, and they might scuttle away from it. On a sunny day it's magical, but then buildings are like that, they should be."[33]

Exemplifying the emotive power and metaphorical potential of architecture, and the possibility of growth as well as decay, a monumental ruin looks to the future as well as the past, generating an appropriate image for a highly regarded and innovative university that is today best known for creative writing and climate change research. But algae, lichen and moss also turned a past image of the future—a new university—into one of slow decay that was appropriate to the demise of free higher education, one of the emblems of the welfare state. After the Second World War, most local authorities paid students' tuition fees and also contributed a maintenance grant towards living costs, which the 1962 Education Act made a legal obligation. But in 1988 the Conservative Education Secretary, Kenneth Baker, signaled a policy change with the quip: "everyone's in debt these days, aren't they?"[34] The first state-supported student loans were for maintenance alone. But Labour's Teaching and Higher Education Act of 1998 introduced tuition fees, which led to a sequence of higher fees and larger loans. In 2003—two years after Lasdun's death—the ziggurats, spine, aerial walkway and other elements of his design were listed Grade II*, initiating a refurbishment program to the specification of English Heritage, the public body responsible for historic buildings. Ignoring that they had become part of the architecture, and were in accordance with the architect's intention, the university appointed a contractor to remove the algae, lichen and moss and apply an anti-fungal inhibitor. The buildings were new again, and the worse for it.

In constructing metaphors and associations, conceiving design and use in relation to time and motion, and imagining UEA's future state, Lasdun emphasized the role of the allegorical imagination in conception and perception. Associating myth-making with designing, he remarked that each architect must devise his or her "own creative myth," which should be "sufficiently objective" and also have "an element of subjectivity; the myth must be partly an expression of the architect's personality and partly of his time, partly a distillation of permanent truths and partly of the ephemerae of the particular moment."[35] Lasdun concluded: "My own myth … engages with history."[36]

While designing UEA, Lasdun was aware of contemporary theories on the classical foundation of modernist architecture, notably Rudolf Wittkower's *Architectural Principles in the Age of Humanism*, 1949, and Colin Rowe's "The Mathematics of the Ideal Villa," 1947.[37] Commenting on contemporary architects who "seek to re-establish the full classical language," Lasdun sympathetically concluded: "They have chosen a hard and difficult road where success is problematic and failure probable."[38] Three centuries before, the Enlightenment's confidence in natural reason had undermined the authority of the classical canon. But rather than wither, classicism had continued as an evolving language open to new influences and mutations. Returning Lasdun's largely positive appraisal of their Hunstanton School, 1954, and recognizing a comparable commitment to history and modernity, Alison and Peter Smithson concluded that Lasdun's success at UEA was the result of "the classical architect's skills … the traditional understanding about size, scale and measure."[39] Peter Smithson had enrolled at the Royal Academy in the late 1940s precisely to acquire a classical expertise, while Lasdun's reassertion of classical principles was latent due to his education. In Lasdun's teens he read J.C. Stobart's *The Glory that was Greece*, 1911, and *The Grandeur that was Rome*, 1912, and studied at the Architectural Association (AA) when the strong Beaux-Arts influence was seen alongside "the early stirrings of modern architecture."[40]

From the Renaissance to the early twentieth century the architect was a historian in the sense that an architectural treatise combined design and history, and a building was expected to manifest the character of the time and knowingly refer to earlier historical eras. Modernism ruptured this system in principle if not always in practice, but it returned with vigor in the mid-twentieth century.

To some degree, mid-twentieth-century modernists merely reaffirmed an appreciation of history that was implicit in works such as Le Corbusier's *Vers une Architecture*, 1923, but largely ignored.[41] In "Mannerism and Modern Architecture," 1950, Rowe quotes Le Corbusier, concluding rather dismissively: "At one moment, architecture is 'the art above all others which achieves a state of Platonic grandeur'; but, at the next, it becomes clear that this state, far from being changeless and eternal, is an excitement subsidiary to the personal perception of 'the masterly, correct and magnificent play of masses brought together in light.'"[42] Concluding that the "whole modern movement appears to share" Le Corbusier's emphasis on subjective experience rather than ideal order, Rowe identifies "the visual index of an acute spiritual and political crisis" in "the present day," which was interwoven with the traumatic consequences of worldwide, military aggression.[43] But the Second World War was a more scientific war than the First, and nuclear devastation undermined confidence in technological progress as a means of social transformation, notably for the generation of architects who had seen military service, such as Lasdun and Peter Smithson.

In the search for stability in the uncertain aftermath of 1945, modernism's previously dismissive reaction to social norms and cultural memories was itself anachronistic. The consequence was not just to acknowledge early modernism's classical heritage but also to place a concern for history at the heart of architecture once again, affirming the liberal humanist tradition that modernism had once seemed to repudiate, and undermining the unnecessary opposition between tradition and innovation that modernism had once seemed to pose. To be ancient and modern was no longer a contradiction. Frances Yates—author of *The Art of Memory*, 1966—described the National Theatre as an "ancient truth in a new idiom," to Lasdun's "enormous pleasure."[44]

Soon after his appointment as UEA's architect, Lasdun told the *Eastern Daily Press* that Nicholas "Hawksmoor was a far greater architect than Frank Lloyd Wright."[45] Later he remarked that Hawksmoor's "point of departure was Ancient Rome but he was convinced that departure from this was essential," concluding that Hawksmoor broke the rules of classical architecture to emphasize them more:[46]

> NH uses elements of classical and gothic (the only elements available to him) in a free, unprejudiced manner not just quoting them but reconstituting them in a new and original whole which is neither classical or gothic but is wholly original and wholly convincing.[47]

If Lasdun had replaced "gothic" for "modern" he could have equally applied this analysis to himself. He was concerned both to emphasize the classical foundation of modern architecture and also to disrupt and transform it in the manner of a modern Hawksmoor.

Architects have used history in different ways, whether to indicate their continuity with the past or their departure from it. But even modernists who denied the relevance of the past relied on histories to validate modernism and articulate its principles. Books such as Pevsner's *Pioneers of the Modern Movement*, 1936, identified a modernist prehistory to justify modernism's historical inevitability, rupture from the past and systematic evolution. Once established, modernism was supposed to remain triumphant. Seemingly contradicting his earlier conjunction of modernism and the picturesque, for which he received fierce criticism, Pevsner remarked in 1961 that at "the beginning of the twentieth century, there arose a generation of giants, who created a new style of architecture, entirely independent of the past" and later concluded: "It seemed folly to think that anybody would wish to abandon [modernism]."[48] But the newly established canon was ripe for revival and reinterpretation, like any earlier architecture. Pevsner took the blame for unwittingly encouraging the "return" of "such a dominating faith in history that it chokes original action and the action which replaces it is inspired by the past."[49] In a BBC radio broadcast later that decade, he contrasted contemporary architecture with his *Pioneers of the Modern Movement*, which was reprinted as

Pioneers of Modern Design in 1949 and revised in 1960. Once again recognizing his "embarrassing" influence on architects who revived styles that he had discarded, Pevsner identified "an anti-Pioneers style … alarmingly harking back to art nouveau and to expressionism."[50] Acknowledging "neo-expressionism" as the more pervasive, he nominated postwar Le Corbusier as its key figure and criticized the Royal College of Physicians, London, 1964, even though he had "the greatest respect for Denys Lasdun," its architect.[51] Lasdun vehemently disliked being called a brutalist because he considered the term to be dogmatic. But Pevsner first denigrated the "over-powering … brutality" of the new style and then, despite his personal distaste, recognized its contemporary relevance as "a successor to my international modern of the 1930s, a postmodern style I would be tempted to call it, but the legitimate style of the 1950s and 1960s."[52] However, he characterized it as a brief interlude associated with the anxious aftermath of war: "phases of so excessively high a pitch of stimulation can't last. We can't, in the long run, live our day-to-day lives in the midst of explosions."[53]

Early Italian modernists had not rejected historical references to the extent of other CIAM members, and Pevsner focused his criticism on postwar Italian architects for furthering this concern, citing Ernesto N. Rogers.[54] Just two years earlier, Reyner Banham had dismissed "the Italian retreat from modern architecture" as "infantile regression."[55] But Lasdun proclaimed:

> If you look at the modern Italian work, for example, it is at least clear that architects have engaged in terrible battle with architecture and certainly many of them have been disastrously defeated; but most English architects seem to have reached a gentlemanly understanding with their art that they should leave each other strictly alone.[56]

Critical of international modernism, Rogers promoted appreciation of national and regional architectural cultures. Advocating "continuity" in 1954, he emphasized: "No work is truly modern which is not genuinely rooted in tradition, while no ancient work has a modern meaning which is not capable of somehow reflecting our modern temper."[57] To explain his conception of a building in dialogue with its physical and natural surroundings and contributing to an evolving historical continuity, Rogers quoted from "Tradition and the Individual Talent," 1917, in which T.S. Eliot emphasizes that the present alters our understanding of the past as much as the past influences the present. Admired by Rogers and equally indebted to Eliot's essay, Lasdun noted the value that the poet placed on innovation as well as tradition: "The existing monuments form an ideal order among themselves, which is modified by the introduction of the new (the really new) work of art among them."[58]

Confirming the prevalence of such ideas in postwar architecture, in 1969 Vincent Scully concluded that the architect will "always be dealing with historical problems—with the past and, a function of the past, with the future. So the architect should be regarded as a kind of physical historian … the architect builds visible history."[59] As a design is a reinterpretation of the past that is meaningful to the present, transforming both, each building or landscape is a new history. The architect is a historian twice over: as a writer and as a designer. We expect a history to be written in words, but it can also be cast in concrete or seeded in soil.

Equating a history to a ruin, Walter Benjamin remarked: "Allegories are, in the realm of thoughts, what ruins are in the realm of things."[60] Our understanding of the past is inevitably partial. Laying bare the processes of construction and decay, a history is both a ruin of the past and a speculative reconstruction. Whether implicit or explicit, a critique of the present and a prospect of the future are also evident in historical statements, which can never be neutral. As a design is equivalent to a history, we may expect the designer as well as the historian "to have a certain quality of *subjectivity*" that is "suited to the objectivity proper to history," as Paul Ricouer concludes.[61] But the designer does not usually construct a history with the rigor expected of a contemporary historian, and we expect the designer to display other qualities of subjectivity as well, whether personal or cultural.

The architectural imagination emphasizes the material and its absence, whether literal or allegorical. Equally, the architect is a "physical historian." In the symbiosis of geography and history in an island nation, these concepts were united in an evolving tradition from the picturesque, to romanticism, to modernism, which chose as its emblem a hybrid of nature and culture that represents potential as well as loss, the future as well as the past. In postwar British architecture, the return of history meant the return of the ruin.

ACKNOWLEDGEMENTS

I very much appreciate the assistance of Roger Bond, Director of Estates & Buildings, UEA; Dominic Bradbury; Professor Peter Brimblecombe, UEA; Dr Barnabas Calder, University of Liverpool; Katherine Clarke, muf architecture/art; Kurt Helfrich, Fiona Orsini and Suzanne Walters, RIBA Drawings & Archives Collections, V & A, London; Kathryn Holeywell, UEA; Christine Hiskey, Archivist, Holkham Hall; James Lasdun; Lady (Susan) Lasdun; and Hal Moggridge, Colvin & Moggridge.

BIBLIOGRAPHY

Banham, Reyner, "Neo-Liberty: The Italian Retreat from Modern Architecture." *The Architectural Review* 125, no. 747 (April 1959): pp. 231–5.

Bates, Stephen, "Tuition Fees: From 'Free' University Education to Students Owing Thousands." *The Guardian* (October 12, 2010). Available at: http://www.guardian.co.uk/education/2010/oct/12/tuition-fees-student-finance-history [accessed August 12, 2012].

Benjamin, Walter, *The Origin of German Tragic Drama*. Translated by John Osborne (London: New Left Books, 1977). Completed in 1928.

Casson, Hugh, Brenda Colvin and Jacques Groag, *Bombed Churches as War Memorials* (Cheam: Architectural Press, 1945).

Colvin, Brenda, *Land and Landscape: Evolution, Design and Control* (London: John Murray, 1970, 2nd edition).

Colvin, Brenda, "A Planting Plan," in Hugh Casson, Brenda Colvin and Jacques Groag, *Bombed Churches as War Memorials* (Cheam: Architectural Press, 1945), pp. 23–30.

Colvin, Brenda, *Wonder in a World*. Privately printed, 1977.

Connell, Brian, "Denys Lasdun: Building a Landscape for Figures," *The Times* (March 24, 1975): p. 10.

Curtis, William, *Denys Lasdun: Architecture, City, Landscape* (London: Phaidon, 1994).

Dalyell, Tam, "Bert Hazell: Trade Union Leader and Labour MP who Championed the Cause of Agricultural Workers," *The Independent on Sunday* 21 (January 2009). Available at: http://www.independent.co.uk/news/obituaries/bert-hazell-trade-union-leader-and-labour-mp-who-championed-the-cause-of-agricultural-workers-1452239.html [accessed December 17, 2011].

Davies, John H.V., and Deny Lasdun, "Thoughts in Progress: Summing Up II," *Architectural Design* 27 (November 1957): pp. 395–6.

Dormer, Peter, and Stefan Muthesius, *Concrete and Open Skies: Architecture at the University of East Anglia* (London: Unicorn, 2001).

Eliot, T.S., *Points of View* (London: Faber and Faber, 1941).

Eliot, T.S., "Tradition and the Individual Talent," in T.S. Eliot, *Points of View* (London: Faber and Faber, 1941), pp. 23–34. First published in 1917.

Lasdun, Denys, "About Anglia: Denys Lasdun Chosen to Design the New Norwich University, 1963". Available at: http://www.eafa.org.uk/catalogue/213000 [accessed June 10, 2012].

Lasdun, Denys (ed.), *Architecture in an Age of Scepticism: A Practitioner's Anthology Compiled by Denys Lasdun* (London: Heinemann, 1984).

Lasdun, Denys, "The Architecture of Urban Landscape," in Denys Lasdun (ed.), *Architecture in an Age of Scepticism: A Practitioner's Anthology Compiled by Denys Lasdun* (London: Heinemann, 1984), pp. 134–59.

Lasdun, Denys, "His Approach to Architecture," *Architectural Design* 35 (June 1965): pp. 271–91.

Lasdun, Denys, "Notes on a Lecture on Nicholas Hawksmoor," in William Curtis, *Denys Lasdun: Architecture, City, Landscape* (London: Phaidon, 1994), p. 223.

Lasdun, Denys, 'Royal Cold Medallist Address," in Curtis, pp. 221–2. First published in 1977.

Lasdun, James, "My Father," *Modern Painters* (Winter 2003): pp. 54–61.

Le Corbusier, *Towards a New Architecture*. Translated by Frederick Etchells (London: Architectural Press, 1946).

Pevsner, Nikolaus, "The Anti-Pioneers," in Pevsner, *Pevsner on Art and Architecture*, pp. 293–307. First published in 1967.

Pevsner, Nikolaus, *The Englishness of English Art: An Expanded and Annotated Version of the Reith Lectures Broadcast in October and November 1955* (London: Architectural Press, 1956).

Pevsner, Nikolaus, "The Genesis of the Picturesque," *The Architectural Review* 96, no. 575 (November 1944): pp. 139–46.

Pevsner, Nikolaus, "The Genius of the Place," in Pevsner, *Pevsner on Art and Architecture*, pp. 230–40.

Pevsner, Nikolaus, "Modern Architecture and the Historian or The Return of Historicism," *The Architectural Review* 68, no. 6 (April 1961), pp. 230–40. Based on a lecture given at the RIBA on January 10, 1961.

Pevsner, Nikolaus, *Pevsner on Art and Architecture: The Radio Talks*, ed. Stephen Games (London: Methuen, 2002).

Pevsner, Nikolaus, "The Return of Historicism," in Pevsner, *Pevsner on Art and Architecture*, pp. 271–8.

Repton, Humphry, *The Landscape Gardening and the Landscape Architecture of the Late Humphry Repton, Esq.*, ed. J.C. Loudon (Farnborough: Gregg International, 1969). First published 1840.

Ricoeur, Paul, *History and Truth*, trans. Charles A. Kelbley (Evanston, IL: Northwestern University Press, 1965).

Ricoeur, Paul, "Objectivity and Subjectivity in History," in Ricouer, *History and Truth*, pp. 21–40.

Robbins, Professor Lord et al., "Committee on Higher Education" (Robbins Report) (London: Her Majesty's Stationery Office, 1963).

Rogers, Ernesto N., "Continuità," *Casabella Continuità*, no. 199 (December 1953–January 1954): p. 2.

Rowe, Colin, "Mannerism and Modern Architecture," in Rowe, *The Mathematics of the Ideal Villa and Other Essays*, pp. 29–51.

Rowe, Colin, "The Mathematics of the Ideal Villa," in Rowe, *The Mathematics of the Ideal Villa and Other Essays*, pp. 1–27.

Rowe, Colin, *The Mathematics of the Ideal Villa and Other Essays* (Cambridge, MA: MIT, 1976).

Sanderson, Michael, *The History of the University of East Anglia Norwich* (London: Hambledon and London, 2002).

Scully, Vincent, *American Architecture and Urbanism* (London: Thames and Hudson, 1969).

Smithson, Alison, and Peter Smithson, *Without Rhetoric: An Architectural Aesthetic 1955–1972* (London: Latimer New Dimensions, 1973).

Stobart, J.C., *The Glory that was Greece* (London: Sedgwick & Jackson, 1911).

Stobart, J.C., *The Grandeur that was Rome* (London: Sedgwick & Jackson, 1912).

Whately, Thomas, *Observations on Modern Gardening, Illustrated by Descriptions* (London: T. Payne, 1771). First published in 1770.

Wittkower, Rudolf, *Architectural Principles in the Age of Humanism* (London: Warburg Institute, 1949).

Woodward, Christopher, *In Ruins* (London: Vintage, 2001).

NOTES

1   Clark, quoted in Christopher Woodward, *In Ruins* (London, 2001), p. 212.

2   Clark et al., in Hugh Casson, Brenda Colvin and Jacques Groag, *Bombed Churches as War Memorials* (Cheam, 1945), p. 4.

3   Brenda Colvin, "A Planting Plan," in ibid., pp. 26, 28, 30.

4   Brenda Colvin, *Land and Landscape: Evolution, Design and Control* (London, 1970, second edition), p. 222.

5   Thomas Whately, *Observations on Modern Gardening, Illustrated by Descriptions* (London, 1771; first published in 1770), p. 131.

6   Colvin, *Land and Landscape*, p. 220.

7   Nikolaus Pevsner, *The Englishness of English Art: An Expanded and Annotated Version of the Reith Lectures Broadcast in October and November 1955* (London, 1956), p. 185.

8   Nikolas Pevsner, "The Genesis of the Picturesque," *The Architectural Review* 96/575 (November 1944): p. 139.

9   Nikolas Pevsner, "The Genius of the Place," in *Pevsner on Art and Architecture: The Radio Talks*, ed. S. Games (London, 2002), p. 232.

10   Professor Lord Robbins et al., "Committee on Higher Education" (Robbins Report) (London, 1963), p. 265.

11   Leicester, quoted in Tam Dalyell, "Bert Hazell: Trade Union Leader and Labour MP who Championed the Cause of Agricultural Workers," *The Independent on Sunday* 21 (January 2009); at http://www.independent.co.uk/news/obituaries/bert-hazell-trade-union-leader-and-labour-mp-who-championed-the-cause-of-agricultural-workers-1452239.html [accessed December 17, 2011].

12   Frank Thistlethwaite, "The Founding of the University of East Anglia: A Reminiscent Chronicle" (November 1963): p. 10. Lasdun Archive, RIBA Library Drawings and Archives Collections, V & A, London.

13   Pevsner, "The Genius of the Place," p. 233.

14   Denys Lasdun, "About Anglia: Denys Lasdun Chosen to Design the New Norwich University, 1963". Available at: http://www.eafa.org.uk/catalogue/213000 [accessed June 10, 2012].

15   Humphry Repton, *The Landscape Gardening and the Landscape Architecture of the Late Humphry Repton, Esq.*, ed. J.C. Loudon (Farnborough, 1969; first published 1840), p. 96.

16   Alexander Redhouse and Peter McKinley, 1966, quoted in Michael Sanderson, *The History of the University of East Anglia Norwich* (London, 2002), p. 147.

17   Denys Lasdun, "His Approach to Architecture," *Architectural Design* 35 (June 1965): p. 273.

18   Lasdun, 1962, quoted in Peter Dormer and Stefan Muthesius, *Concrete and Open Skies: Architecture at the University of East Anglia* (London, 2001), p. 156.

19   Denys Lasdun, "The Architecture of Urban Landscape," in Denys Lasdun, ed., *Architecture in an Age of Scepticism: A Practitioner's Anthology Compiled by Denys Lasdun* (London, 1984), p. 135.

20   Brenda Colvin, *Wonder in a World* (1977), p. 2.

21   Moggridge, in conversation with Hill, November 11, 2013.

22   Brenda Colvin, "Interim Landscape Report and Approximate Estimate of Cost, UEA" (December 1967): pp. 1–2, 21; Colvin & Moggridge Archive.

23   Moggridge, in conversation with Hill, November 11, 2013.

24   Lasdun, "The Architecture of Urban Landscape," p. 146; Lasdun quoted in William Curtis, *Denys Lasdun: Architecture, City, Landscape* (London, 1994), p. 96; Lasdun, "His Approach to Architecture," p. 273.

25   Colvin, "Interim Landscape Report and Approximate Estimate of Cost, UEA," p. 17, dwg. 511/R/5. Colvin & Moggridge Archive.

26   John H.V. Davies and Deny Lasdun, "Thoughts in Progress: Summing Up II," *Architectural Design* 27 (December 1957): p. 395.

27 The photographer is Richard Einzig.

28 James Lasdun, "My Father," *Modern Painters* (Winter 2003): p. 58.

29 Lasdun, quoted in Dormer and Muthesius, p. 70.

30 Kathryn Holeywell, in email to Hill, December 22, 2011.

31 Lady (Susan) Lasdun, in conversation with Hill, August 16, 2012.

32 Lasdun, in Simon Jenkins, "Interview with Denys Lasdun," April 19, 1979, Lasdun Archive; Lasdun quoted in Brian Connell, "Denys Lasdun: Building a Landscape for Figures," *The Times* (March 24, 1975): p. 10.

33 Lasdun, in "Interview with Denys Lasdun," Revised Draft, June 13, 1979, p. 11; Lasdun Archive.

34 Baker, quoted in Stephen Bates, "Tuition Fees: From 'Free' University Education to Students Owing Thousands," *The Guardian* (October 12, 2010). Available at: http://www.guardian.co.uk/education/2010/oct/12/tuition-fees-student-finance-history [accessed August 12, 2012].

35 Lasdun, "The Architecture of Urban Landscape," p. 137.

36 Ibid., p. 139.

37 Lasdun praised Wittkower in Davies and Lasdun, p. 436; Rudolf Wittkower, *Architectural Principles in the Age of Humanism* (London, 1949), p. 135; Colin Rowe, "The Mathematics of the Ideal Villa," in Colin Rowe, *The Mathematics of the Ideal Villa and Other Essays* (Cambridge, MA, 1976), pp. 2–14.

38 Lasdun, "The Architecture of Urban Landscape," p. 137.

39 Alison Smithson and Peter Smithson, *Without Rhetoric: An Architectural Aesthetic 1955–1972* (London, 1973), p. 30.

40 Denys Lasdun, "Royal Gold Medallist Address," in Curtis, p. 221.

41 Le Corbusier, *Towards a New Architecture*, trans. F. Etchells (London, 1946), pp. 31, 124, 125, 130, 135.

42 Rowe quotes from the 1927 edition of *Towards a New Architecture*; the pagination is the same in the 1946 edition. Le Corbusier, *Towards a New Architecture*, 102, 31, quoted in Colin Rowe, "Mannerism and Modern Architecture," in Rowe, *The Mathematics of the Ideal Villa and Other Essays*, p. 42.

43 Rowe relates sixteenth-century mannerism to twentieth-century modernism. Rowe, "Mannerism and Modern Architecture," p. 42–3.

44 Yates, letter to Lasdun, May 17, 1976; Lasdun, letter to Yates, May 21, 1976, Lasdun Archive.

45 Denys Lasdun, March 3, 1962, quoted in Dormer and Muthesius, p. 156.

46 Denys Lasdun, "Notes on a Lecture on Nicholas Hawksmoor," in Curtis, p. 223.

47 Ibid.

48 Nikolas Pevsner, "Modern Architecture and the Historian or The Return of Historicism," *The Architectural Review* 68, no. 6 (April 1961), p. 230; Pevsner, "The Anti-Pioneers," in Pevsner, *Pevsner on Art and Architecture*, p. 295.

49 Nikolas Pevsner, "The Return of Historicism," in Pevsner, *Pevsner on Art and Architecture*, p. 271. Refer to Pevsner, 'Modern Architecture and the Historian,' p. 230.

50   Pevsner, "The Anti-Pioneers," p. 295, 305.

51   Ibid., p. 298.

52   Ibid., p. 299.

53   Ibid., p. 307.

54   Pevsner, "Modern Architecture and the Historian," pp. 231–3.

55   Reyner Banham, "Neo-Liberty: The Italian Retreat from Modern Architecture," *The Architectural Review* 125/747 (April 1959): p. 235.

56   Lasdun, in Davies and Lasdun, p. 436.

57   Ernesto N. Rogers, "Continuità," *Casabella Continuità* 199 (December 1953–January 1954): p. 2.

58   T.S. Eliot, "Tradition and the Individual Talent," in T.S. Eliot, *Points of View* (London, 1941; first published in 1917), pp. 23–34.

59   Vincent Scully, *American Architecture and Urbanism* (London, 1969), p. 257.

60   Walter Benjamin, *The Origin of German Tragic Drama*, trans. John Osborne (London, 1977), p. 178.

61   Paul Ricoeur, "Objectivity and Subjectivity in History," in *History and Truth*, trans. C.A. Kelbley (Evanston, IL, 1965), p. 22.

13

Adventures in Angelic Material Imagination:
The Baroque and the Digital as Recounted by Putto_1435

Alessandro Ayuso

BODY AGENTS AND THE MATERIAL IMAGINATION

"Body agents" are figural representations enmeshed in architectural constructions. Their placement, gestures and attributes catalyze connections between potentially disparate variables in a given architectural situation. Through their recurring appearances in drawings, and sometimes through their personification in narrative text, they form evolving relationships with other entities in a design scenario, a process of constructing inter-subjectivities.[1] Body agents can "initiate an adventure in perception," one grounded in the ambiguous dialectic between the reality and the unreality principle.[2] The presentation of figures that appeal to empathetic, individual and corporeal sensibilities transforms everyday, mundane matter in the imagination.

To put forth the argument that body agents activate the material imagination as part of a crucial performance in Baroque architecture, this chapter examines three *putti* designed by Gian Lorenzo Bernini in St. Peter's Basilica and explores how, as body agents, they may continue to serve a role in contemporary design practices for an ongoing project. Architects are increasingly confronted by the ambiguity of a "new materiality,"[3] and a reconsideration of the "conceptual technology" of Baroque body agents—which aided an architect's imagination by incorporating sensation, movement and materiality in a cohesive metaphysical project—heightens the possibilities for the material imagination to play an expanded role in today's designs.[4]

Taking Virginia Woolf's *Orlando* as a literary model, this chapter incorporates a collection of fictional narratives from the viewpoint of a 579-year-old body agent who has witnessed the historic Baroque era—referred to here as "Putto_1435." He recounts his participation in Bernini's interconnected "solar system" of projects at St. Peter's Basilica, and in a contemporary design project.[5] Putto_1435 is "evoked" throughout history, and in each instance takes on a slightly altered form. Having witnessed so many eras first hand he is historically erudite, yet he is histrionic, playful, egotistical, and obsessed with memories of "lost opportunities" and "missed roles" in the work of Michelangelo. The latter characteristics tend to make him an "unreliable narrator." As Henry James writes in his preface to *The Princess Casamassima*, "the figures in any picture, the agents in any drama, are interesting only in proportion as they feel their respective situations," creating a level of empathy that achieves a "finely aware and richly responsible" level of consciousness.[6] Here, the act of writing allows for interpretations of the "feeling" of the Putto_1435 in Bernini's sketches and serial representations, and as a generative aspect to the design of contemporary projects.

BODY AGENT *PUTTI*

Putti are an ancient Roman form of ornament consisting of male toddlers, usually nude, sometimes depicted with wings, that was revived during the *quattrocento* in Italy. In Roman antiquity, *putti* could be found as ornaments on funerary monuments mourning the death of a child, but also in illustrations of Bacchanalia.[7] Their appearances diversified and evolved during the *quattrocento* and the centuries that followed.[8]

The vast interior space of St. Peter's Basilica is punctuated by a multitude of figural ornaments. The *putti* decorating the "thick 2D" of the interior surface, from the fonts of holy water at the lowest register of the sanctuary to the uppermost reaches of the dome over the crossing, make for a non-Cartesian spatial constellation.[9] While adhering to typological constraints, Bernini's *putti* often have unique individual qualities.[10]

Bernini's *putti* are productively ambiguous in three ways. Firstly, as angelic beings they possess traits and abilities that appear as "other than"—divine entities apart from quotidian existence—yet they are repositories and expressions of "techniques of the body," and evoke familiarity.[11] Secondly, their spatial positioning is often liminal—situated along borders and edges of niches, frames and zones, they stitch these disparate parts together. Finally, their designs express varying degrees of adherence to empirically verifiable physicality: sometimes they retain the outward appearance of material and corporeal stability, and at other times they demonstrate corporeality and materiality that are blurred or unstable.

BALDACCHINO (1624–1633): TECTONIC UNREALITY

Design Process

Let me tell you about my involvement with Signor Bernini. To tell you this story is to give you a virtual timeline of his career; I appeared in virtually every one of his finished architectural works! I first appeared in 1624 in the Baldacchino, directly under the magnificent duomo designed by Il Divino.[12]

This progetto was not only Signor Bernini's; the design took place over many years, and there were many hands

involved. Signor Maderno first sketched my companions
dancing in the vines that grew up the twisted columns.[13]
You may miss seeing them there if you visit, because
they are merely the tiniest of reliefs.[14] Unlike their
insignificant parts, my appearance in the project is one
of my finest and most glorious starring roles, and you
will most certainly not miss it if you walk into that
splendid basilica!

As Signor Bernini sketched the design, my companions and
I appeared and disappeared, trying out different poses
and jobs. We perched on the pinnacle of la corona, and
when we did, it extended upwards to give us a better
view. In a few of the sketches, we helped to stretch it
further upward by lifting the finial orb at its top.
Then we were feeling quite tired from all the running
about and hoisting, so Signor Bernini drew us as we were
lying on the volute supports, and they bowed gently under
our weight, much in the manner that Il Divino had drawn
the old men and women and their tetti spezzati in the
Sagrestia Nuova.[15]

Bernini explored the structure and ornamentation of the roof canopy through sketching.[16] Angels and angelic *putti* are depicted, alternately "carrying" and embellishing the connection of the serpentine superstructure of the canopy and the Solomonic columns (Figure 13.1).[17] As in Michelangelo's elevations and sketches of sculptural ornaments and voluted pediments for tombs for the New Sacristy, changes in the figures' actions, postures and positions correlate with the form and geometry of the scroll-like forms, suggesting that the figure is a vehicle encountering the design at a pliable stage, aiding the designer to sense its material qualities in absentia. The figures touch the surfaces of the imagined design, likely initiating a thought of the temperature and texture of the material. They lie and pull on the structure's volutes and garlands; the structural volutes compress when the *putti* are shown lying down on their center-points, changing to a more vertical, tall configuration when the *putti* are drawn at their pinnacles in upright positions.

Built Work

In the final work my job is the best: Qua, I stand
triumphantly with my companions on the canopy of the
Baldacchino, 20 braccia in the air, five times larger
than my companions below, right in the center of
everything, displaying the papal tiara![18]

Alora, reader, I can admit to you that I was slightly
scared, holding that corona preziosa just over the lip
of that tall Baldacchino (which, in fact was much taller
than the Cavalcanti Altar), but I resisted the temptation
to show my fear—for I knew that over me, in fact high
above me in the pennacchi, swinging on swag from the
lower part of the drum, and in the lantern of the duomo
itself, were other putti, and they all appeared playfully

13.1   *Baldacchino superstructure study*, by Gian Lorenzo Bernini.

at ease—so I proudly thrust that splendid artifact
forward![19]

The view from here is one of the best that I have ever
been privileged to witness. I am bathed in the light that
streams down on me from the windows in the drum of the
cupola above. I float in this heavenly light day in and
day out, as pilgrims and visitors stream in through the
entry of the Basilica; I smile confidently down on them
as they crane their necks with awe and wonder.

Even though in some sketches, it appears that the *putti* may be aiding the support of the canopy structure, Bernini relieved his *putti* of "heavy lifting" tasks in the Baldacchino.[20] It is the angels standing atop the columns that "lift" the roof; they appear to effortlessly grasp vines that spring forth from the brackets with only their thumbs and index fingers, and with their pinkies daintily arched upward.

The two *putti* positioned above the center of the Baldacchino's cornice make for a prominent focal point within the spatial volume of St. Peter's. In this instance, they are cast from bronze, but unlike the "tiny" *putti* in the columns below, they are not in relief. Instead, they are released from their mooring to the larger form. One of them, Putto_1435, even appears to "float above" the canopy's wooden cornice, discreetly attached to the structure with iron tie-rods from the back.[21] This *putto's* prominently displayed levitation aids in defining an illusion of an anti-gravity zone, and the possibility, particularly from the viewpoint of a worshipper approaching the crossing, of a canopy supported by angels. The *putti* and angels work together to allow a supersession of the apparent physical connection between roof and column by an "unreality principle"; they invite perceptive visitors to take an imaginative leap.

URBINO TOMB (1627–1647): EVERYDAY *PUTTO*

Design Process

In his sketches of the Tomba di Urbino VIII di Roma,
Signor Bernini depicted me standing next to the effigy of
Carità. I realized that the tomb design was very similar
to those in the final work of Il Divino's Sacristy, in
which I ultimately never appeared.[22] It looked like I
would have the chance once again to frolic and tumble
from a tetto spezzato, and that maybe this time I could
appear in stone atop it![23]

In his first pen-and-ink wash drawing, I nestled in the
folds of Carità's drapery, and my companion suckled on
her breast.[24] All seemed well. But then Signor Bernini
made the first clay bozzetto. My companion, with whom
I had hovered atop the Baldacchino, was on the opposite
side of the tomb, relaxing in the shadows of a perfectly
shaped hollow created by Giustizia's draped tunic; when I
tried to do the same, Carità shooed me away!

13.2   *Charity with Four Children* (study for tomb of Pope Urban VIII), by Gian Lorenzo Bernini, c. 1627–34. Musei Vaticani, Vatican City.

```
Signor Bernini made another bozzetto, and this time
two other companions appeared, and Carità, distracted,
stopped paying attention to me (Figure 13.2). Then
Carità nudged me much harder, and she even scolded
me! Such a seemingly gentle maiden with a breast
tantalizingly exposed—how could she do such a coarse
thing, and with such disproportionate strength? To add
to my bewilderment, Carità looked on with amusement as
I reeled. It wasn't just Carità's actions that made me
lurch with such vehemence: while Signor Bernini made the
bozzetto, he would insert it into a wooden model of the
architecture, and I was squeezed by this, to the point
that it left its indentation on my flesh. Dear reader, at
this point I went from put-off to extremely agitated, I
must admit.²⁵
```

In the case of the tomb, the architectural surround was already complete, and the broad strokes of the composition, an elaboration on a standard tomb form persisting since the Renaissance, was already established early on in the pen-and-ink sketch. The interplay between the figures evolved further via the sequence of malleable *bozzetti*, or clay sketch models, which were likely periodically placed in a wooden model of the architectural surround.[26] The design media, observed in sequence, show a progressing intensification of Putto_1435's expression. Putto_1435 is in between the lower vertex of a pyramidal composition of the tomb and the relative mass of the column, which expresses the structural load of the semi-hemispherical apse above (Figure 13.3) and is at the optimal location to express the implicit squeeze between massive architectural surround and sculptural elements.

Built Work

```
When Il Divino had sketched us in the heap below the
tetto spezzato, it was delightful fun, but in this case
I wasn't able to play any games. Qui, deprived of my
ability to fly, I am wedged in between Carità and the
column behind me. Furthermore, atop the tetto spezzato—
even if I could get to it—there were buzzing bees!²⁷ I
know my companions who appeared in the twisted columns
of Il Baldacchino with me were not daunted by these
loathsome (and to my envy, winged) arthropods—but this
member of the animal kingdom was new to me, and insects
in general are not to my liking, and certainly not those
who can inflict harm! Dear reader, the buzzing of these
foul and dangerous creatures echoed throughout the niche
as they clung and flitted about the tomb; but there
was an even more dreadful morte secca that I noticed
Signor Bernini sketching, and who I unfortunately was
to see more of later: an ominous skeleton, flapping his
scraggly wings and dedicating himself, back turned to the
spectators, to scrawling in his devilish book. This shook
me to the core. I longed to fly far from such ghouls!²⁸
Additionally, there were in my peripheral vision, huge,
```

> looming figures. Papa Urbino and the Doctors of the
> Church towered above us all, gesturing grandly.[29]
>
> Reader I tell you, I felt scared, trapped, angry,
> rejected and not least, squeezed. I did not like this
> position one bit, in between towering, malicious or
> indifferent figures, the remains of the deceased, and
> huge columns.

As the tomb was built, in a flanking niche in the apse at St. Peter's, Putto_1435 is an empathetic access point, preparing the congregant to engage emotionally with the architecture and ornament above. This is achieved in several ways. One factor is that Putto_1435 appears less like an angel, and more like a mortal: he is wingless, and wears an unbridled expression of grief. Another factor is Putto_1435's materiality: he is carved from stone identical to that comprising the accompanying *putti* and Charity, and nearly identical to the Corinthian column that appears to squeeze against him. This stereotomic mass differs starkly from the dark bronze and gilt gold effigies of the pope and the personification of death, which are hollow inside, shells left by the lost-wax casting method from which they were made. He is also at the lowest spatial strata of the Basilica's apse, nearly at eye level with seated congregants.

PETRI GLORIA (1656–1666): COMMINGLING WITH THE CLOUDS

Design Process

> As Signor Bernini sketched the clouds and rays of
> light that would surround this luminescent aperture,
> I moved towards them. I wanted to be far from the
> ghoulish figures below, and I wanted to be closer
> to the comfortable sunlight that I basked in atop Il
> Baldacchino.
>
> I hopped from cloud to cloud; I began to find that,
> when I reached out to touch them, my body was fusing
> with them! More putti joined me, and as we celebrated,
> we merged with these atmospheric condensations. It soon
> became difficult to distinguish who was who, and what was
> cloud and what was putto.

Of the three examples of Bernini's work discussed here, the Gloria surround shows the most extreme examples of liberty taken to the expressive materiality and body image of the designed *putti*.[30] A series of Bernini's sketches formed a testing ground where he explored what could be achieved with three variables: light (as rod-like rays), *putti*, and clouds. He experimented with the *putti*'s gestures, from triumphantly flying upwards with backs arched, to crouched in pious reverence.

The *putti* multiply and their bodies become more fluid as the design progresses. In one sketch, a *putto* is atop a cloud outlined in deliberately reinforced contours. Here, the ethereal substance of the cloud supports a body that appears just as

13.3   View showing Cathedra Petri and Gloria window, by Gian Lorenzo Bernini. Personification of Charity and "Putto_1435" of the Tomb of Urban VIII, also by Gian Lorenzo Bernini visible in the lower right hand corner.

fleshy and substantial as any other.[31] But another drawing shows a *putto* with an even more angelic relationship towards gravity and mass; he does not appear to be supported by the clouds at all, but rather appears just as buoyant and immaterial as the clouds themselves.[32] A cherubic face emerging from the cloud in the same sketch suggests yet further ambiguous corporeality: here, the *putto* is literally part of them. It is clear that the dynamic mass of roiling *putti*–cloud hybrids and the surrounding "rays" of light accumulated expansive momentum during the process of sketching. Taking these drawings in chronologic sequence, the atmospheric mass appears to expand explosively, enveloping the window and then eclipsing the boundaries of the niche.[33] The divinely unstable microclimate transgresses the architectural boundaries that could confine it.

Built Work

```
This position is such a vast improvement from my previous
circumstances. From where I stand now peeking out from
behind a bundle of light rays, I truly have the best
view of all. I can see across the entire expanse of the
interior. The evening light below me warms my feet. I
look down and I see it streaming through the window,
reflecting from the putti-clouds around it.
```

The copious images of blurred clouds and *putti* were sculpted from plaster; all the exposed surfaces were gilt a reflective gold (Figure 13.3). The bodies of many of the *putti* in the Gloria surround, physically fluid and contingent, are "other than," and show the proximity of divine forces signified by the dove in the stained glass window's graphic. Similar to Bernini's partially disembodied *putti* in the Albertoni Chapel, also sculpted from plaster, they reflect the light from the nearby window and direct the visitor's attention to the primary focus, the window itself, through the direction of their rapt gazes. While Putto_1435's appearance in the Urban Tomb seems to blend with the stone structure, the gilt plaster of the Gloria surround blends with the light of the window. Through the expression of the materiality of the body, a distinction is made between the earthly and divine realms.

THE BODY AGENT AEDICULA (2014): DIGITAL MATERIALITY

```
But, reader, that was all hundreds of years ago. Since
then, I have contributed to many unusual scenarios,
appearing in Rococo designs in Austria and Bavaria, and
even in Mexican Churrigueresque retablos (these were
great fun). I had many roles as cupid, which I would say
were most boring; and then: nothing. I had no work for
a century of Modernism, during which time I sat in the
clerestory of the apse of San Pietro in Montorio, nursing
bottles of vino santo and, well, feeling a bit depressed.

Now, I have been evoked again, and what I have undergone
in the present day was, at first, certainly most
bewildering. Often, in the new setting where I find that
design often takes place, there is no ground at all;
and if there is a ground, it reminds me of those grids
carefully drawn by Leonardo, receding infinitely into
the distance.³⁴ My skin often has a similar appearance
to the grid as if it had been overlaid onto my body. And
disturbingly, I am hollow inside, save for my teeth and
tongue, and the two floating orbs that are my new eyes.
My flesh materializes and dematerializes sporadically;
when the dematerialization occurs, I can be seen through
and through my carapace.³⁵

I am slowly growing accustomed to this odd form that
I have taken. Recently I assisted in the design of un
```

piccolo edificio. I thought that it could be dedicated to
the worship of Il Padre, although I don't think this is
the intent.[36]

The design began with many sketches done with pencil and
paper. This process was, of course, familiar to me. But
then the disegno proceeded through a series of films made
in the gridded space.[37]

I began to take part in the films. Again I found myself
with my companions, flying around the top of what was
becoming a tempietto. I spiraled overhead, but also
walked through it. I found folds, shelves, and niches
that were forming (I am always attracted to such spaces).
I played games with the other figures (they seemed quite
unsuspecting), hiding behind corners and surprising
them as they innocently walked through the convoluted
tempietto. It was a delight to watch their startled
reactions! There were times that I became entangled
in the folded surface and I found I could leave my
impression in it.[38] I touched the surfaces and listened to
the echoes of sound that ricocheted off of the planes. At
first, there were only a few planes, large and flat, but
as we walked through them and touched them, they became
smaller and the forms softer, offering niches to hide
within, much like Giustizia's gown that I so envied as a
hiding spot in the design of Papa Urbino's tomb.

One thing that I enjoyed most about working with Signor
Bernini was that he often encouraged me to fly upwards.[39]
So here, in this very free environment of digital space,
I did not hesitate to move myself to the top of what
seemed like a combination of a duomo piccolo and an oculo
alto. Even though it does not look exactly like those
that I am used to, I know how it will work and what I
should do. When visitors enter this portion of this nuovo
tempietto, they will surely look up towards the light
drifting down to them at the top; and I will be there
too, looking down at them![40]

My next appearance was in the construction of a modello.
The stuff that the modello is made from is strange
indeed; it is certainly unlike clay. It is a strange
white powder, light and fine, deposited layer by layer
by a machine. It takes any form that I help to make.
As I balanced at the rim of oculus, I kneaded my hands
into it and I found they merged with it; this reminded
me of when I merged so seamlessly with the clouds of the
Gloria window. As my hands fused with the tempietto, I
nearly lost my balance again, and additional supports
sprouted from my back. In some ways, this reminds me
of Il Baldacchino, when reinforcements supported me on
that canopy, but unlike Il Baldacchino these new ones

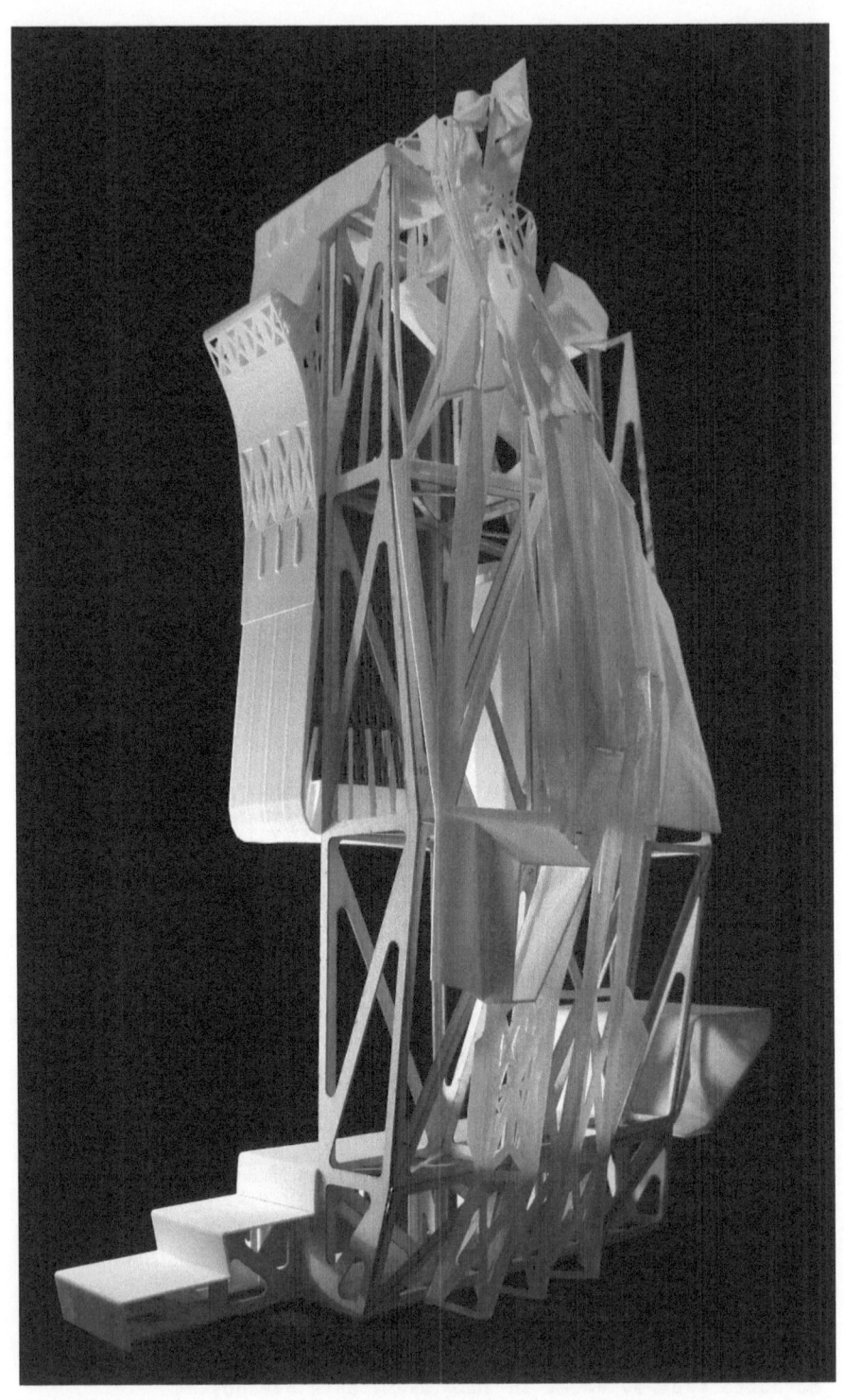

13.4　View of Body Agent Aedicula 1:10 sectional model, by Alessandro Ayuso.

```
are not hidden but rather, they are exposed. To me,
they are splendid ornaments, a new set of accoutrements
to complement my new wings, which were designed and
materialized in this strange way out of this special
substance.
```

In the "films" that Putto_1435 refers to, the design of a freestanding architectural installation developed in an evolving, digitally produced animation. Sequences of body agents' movements and actions were choreographed and adjusted as the design progressed, visualized through deliberate framing and camera tracking. In the animation, Putto_1435's performance was both operative and evocative. The operative function arose from the deflection of attractor points embedded in the mesh planes; because of these, when the body agents moved, the meshes would distort correspondingly; these planes gradually transformed to define a spatial volume. The evocative aspects of Putto_1435's actions stemmed from the cinematic quality of the animations, which incorporated his point of view and his dynamic image. The visualizations were intensified by the simultaneous composition of Putto_1435's narrative.

13.5   Detail view of upper portion of Body Agent Aedicula 1:10 sectional model, two ornamental *putti*, by Alessandro Ayuso.

His biography compiles a vivid set of practices forming an intensely hybrid habitus.[41] It situates the imagination of his contact with the neutral, malleable planes from the reference point of a specific corporeality.

In the 3D printed model that followed, Putto_1435's body merges with the superstructure of the installation. Fabricated from a single Selective Laser Sinter build, his body is continuous with the larger structure he adorns. The fused nylon powder that results from the SLS process, a corollary of the digital design image, is without overt, sensual qualities, but Putto_1435's presence is an insistent catalyst to adventurously explore and imagine possible relationships between particular bodies and new materialities.

CONCLUSION

During Bernini's design process, body agents were proxies that allowed him to sense the architecture in absentia. In the Baroque imagination, materiality oscillates between fixed forms and utter pliability.[42] The body agent operates in between the two systems, bringing empathetic imagination to bear on the malleable forms, and aiding in the imagination of possible material crystallizations. As built ornament, the figure continues to mediate ambiguous circumstances, but in Bernini's work at St. Peter's it aids in the negotiation of the borders between physical adjacencies as well as between the world of the divine and quotidian. Through their expression, placement, and materiality the body agents make appeals directly to the visitors of St. Peter's, guiding their reveries. In the narrative, having recalled his participation in Bernini's projects, where his body morphed and even at times seemed to fleetingly evaporate, Putto_1435 becomes more comfortable with the exuberant transmutability of material and form in the digital realm. With their specific, empathetic and catalytic qualities, body agents aid designers' material imaginations: their performance bypasses qualitative neutrality by encouraging wonder in an adventure in perception.

BIBLIOGRAPHY

Allen, Stan, "Mat Urbanism: The Thick 2-D," in Hashim Sarkis (ed.), *CASE: Le Corbusier's Venice Hospital* (Munich: Pastel/Harvard Design School, 2002), pp. 118–26.

Bachelard, Gaston, *Earth and the Reveries of Will: An Essay on the Imagination of Matter*, trans. Kenneth Haltman (Dallas, TX: Dallas Institute Publications, 2002).

Balas, Edith, *Michelangelo's Medici Chapel: A New Interpretation* (Philadelphia, PA: American Philosophical Society, 1995).

Careri, Giovanni, *Bernini: Flights of Love, the Art of Devotion* (Chicago, IL: University of Chicago Press, 1995).

Dempsey, Charles, *Inventing the Renaissance Putto* (Chapel Hill, NC: University of North Carolina Press, 2001).

Dickerson, C.D. and Sigel, Anthony, "Charity with Four Children," in Nancy Grubb (ed.), *Bernini: Sculpting in Clay* (New York, NY: Metropolitan Museum of Art, 2012), pp. 112–17.

Harbison, Robert, *Reflections on Baroque* (Chicago, IL: University of Chicago, 2001).

Hersey, George, *Architecture and Geometry in the Age of the Baroque* (Chicago, IL: University of Chicago Press, 2001).

Hills, Helen, "Introduction," in Helen Hills (ed.), *Rethinking the Baroque* (Farnham: Ashgate, 2011).

James, Henry, "Preface to *The Princess Casamassima*," in *The Art of the Novel: Critical Prefaces* (Chicago, IL: University of Chicago Press, 2011), pp. 59–78.

Lavin, Irving, "The Baldacchino. Borromini vs. Bernini: Did Borromini Forget Himself?", in Georg Satzinger and Sebastian Schütze (eds), *St. Peter in Rom 1506–2006: Akten der internationalen Tagung 22.–25.02.2006* (Munich: Hirmer Verlag, 2008), pp. 275–300.

Lavin, Irving, *Drawings by GianLorenzo Bernini* (Princeton, NJ: Art Museum, Princeton University, 1981).

Marder, T.A., *Bernini and the Art of Architecture* (New York, NY: Abbeville Press Inc, 1998).

Mauss, Marcel, "Techniques of the Body," in Jonathan Crary and Stanford Kwinter (eds), *Incorporations* (New York, NY: Zone, 1992), pp. 455–77.

Picon, Antoine, *Digital Culture in Architecture* (Basel: Birkhauser, 2010).

Rossholm Lagerlöf, Margaretha, "The Apparition of Faith: The Performative Meaning of Gian Lorenzo Bernini's Decoration for the Cornaro Chapel," in Peter Gillgren and Marten Snickare (eds), *Performativity and Performance in Baroque Rome* (Burlington, VT: Ashgate, 2012), pp. 179–80.

Spuybroek, Lars, *The Sympathy of Things: Ruskin and the Ecology of Design* (Rotterdam: V2, 2011).

van Toorn, Roemer, *The Quasi Object: Purity and Provocation in the Library of Utrecht by Weil Arets*. Available at: http://www.roemervantoorn.nl/quasiobject.html [accessed February 20, 2014].

NOTES

1   The construction of inter-subjectivities in architecture is also a quality of the "quasi-object." Roemer van Toorn, *The Quasi Object: Purity and Provocation in the Library of Utrecht by Weil Arets*. Available at: http://www.roemervantoorn.nl/quasiobject.html [accessed February 20, 2014].

2   "[T]he real stands before us in all its terrestrial materiality … we are easily persuaded that the reality principle must usurp the unreality principle, forgetting the unconscious impulses, the oneiric forces which flow unceasingly through our conscious life." Gaston Bachelard, *Earth and the Reveries of Will: An Essay On the Imagination of Matter*, trans. K. Haltman (Dallas, 2002), p. 3.

3   "The new materiality is located at the intersection of two seemingly opposed categories, the totally abstract, based on signals and codes, and the ultra-concrete, involving an acute and almost pathological perception of material phenomena and properties." Antoine Picon, *Digital Culture in Architecture* (Basel, 2010), p. 157.

4   On the topic of "conceptual technology," Helen Hills writes, "some scholars … have explored baroque … as a 'conceptual technology' that does not simply allow retrospective understanding but actually provokes new forms of historical conceptualization and interpretation." Helen Hills, "Introduction," in Helen Hills (ed.), *Rethinking the Baroque* (Farnham, 2011), p. 3. On the topic of sensation and movement in the Baroque, Robert Harbison writes, "In the Baroque, artists began to think of creating sensation …." Elsewhere, he claims, "The Baroque sets itself apart from what precedes it by an interest in movement above all … [this] is not just a formal exercise but signals a transformation of the human consciousness." Robert Harbison, *Reflections on Baroque* (Chicago, 2001), pp. 2 and 225. The metaphysical approach to materiality in the Baroque is described by George Hersey, *Architecture and Geometry in the Age of the Baroque* (Chicago, 2001), pp. 18–21.

5   "[U]nlike his predecessors at St. Peter's, Bernini did not conceive of the Baldacchino ideologically as an isolated monument, but the focal point of a veritable solar system of memorabilia …." Irving Lavin, "The Baldacchino. Borromini vs. Bernini: Did Borromini Forget Himself?", in Georg Satzinger and Sebastian Schütze (eds), *St. Peter in Rom 1506–2006: Akten der internationalen Tagung 22.–25.02.2006* (Munich, 2008), p. 289.

6   Henry James, "Preface to The Princess Casamassima," in *The Art of the Novel: Critical Prefaces* (Chicago, 2011), p. 62.

7   An Example of Bacchic imagery on a Roman sarcophagus can be seen in the Roman Bacchic sarcophagus of the second century A.D. in Pisa, Campo Santo.

8   For extensive discussion on *putti* in the *quattrocento* era, see Charles Dempsey, *Inventing the Renaissance Putto* (Chapel Hill, 2001).

9   Although the term "thick 2D" was originally used by Stan Allen with regards to landscape urbanism and mat buildings, the "compact and highly differentiated section" that is "not the product of stacking (discrete layers as in a conventional building section) but of weaving, warping, folding, oozing, interlacing, or knotting together" is an apt description of the walls of St. Peter's where in a given wall section one is likely to find a complex interweaving of ornament, sculpture, load-bearing material and inhabitable space within the wall's pochè. Stan Allen, "Mat Urbanism: The Thick 2-D," in Hashim Sarkis (ed.), *CASE: Le Corbusier's Venice Hospital* (Munich, 2002), p. 125.

10   Bernini sought the individual in each figure. "[His] ambition to make every figure individual is an artistic strategy with a Christian paradigm." Margaretha Rossholm Lagerlöf, "The Apparition of Faith: The Performative Meaning of Gian Lorenzo Bernini's Decoration for the Cornaro Chapel," in Peter Gillgren and Marten Snickare (eds), *Performativity and Performance in Baroque Rome* (Burlington, 2012), pp. 179–80.

11   Body techniques refer to "the ways in which, from society to society, men know how to use their bodies." Marcel Mauss, "Techniques of the Body," in Jonathan Crary and Stanford Kwinter (eds), *Incorporations* (New York, 1992), p. 455.

12   Here, as in many other places, Putto_1435 also claims to "have worked with" Michelangelo Buonarroti, and refers to him as "Il Divino."

13   Carlo Maderno was a Swiss–Italian architect who was responsible for several designs at St. Peter's Basilica, including the façade and the initial schemes for the Baldacchino, before Bernini's involvement with the design.

14   Putto_1435 refers to other *putti* and ornamental figures that he appears in works with as his companions.

15   In Bernini's sketch featuring recumbent *putti*, the *putti* are in poses similar to the effigies known as Night, Day, Dawn and Dusk. Michelangelo's "agent in Carrera ... [referred] to the female allegories as 'the nude figures'; elsewhere, they are spoken of as 'the two women', and their male companions are simply the 'old men.'" They are positioned atop the broken pediments of the Ducal tombs in the New Sacristy in Florence, date 1521–55. Edith Balas, *Michelangelo's Medici Chapel: A New Interpretation* (Philadelphia, 1995), p. 6.

16   Since Bernini was charged with designing a *baldachin*, a temporary structure with a fabric roof detached from staves, rather than a more permanent *ciborium*, he struggled to achieve the expression of the *baldachin*, when in fact, the columns did support the heavy roof. The ornamental figures were part of the articulation of the *baldachin* type. This is described in T.A. Marder, *Bernini and the Art of Architecture* (New York, 1998), pp. 27–46.

17   Also see *Baldacchino study*, by Gian Lorenzo Bernini. Red chalk and pen over black chalk, 14 ½ × 10 ¼ in. (36 × 26.3 cm), Graphische Sammlung, Albertina, Vienna and Baldicchino canopy study by Gian Lorenzo Bernini. Pen on paper, c. 1631. Biblioteca Apostolica Vatican, Barb. Lat. 9900, fol. 2.

18   In Renaissance Italy, 1 braccio = roughly 0.7 meters.

19   Vasari noted that the *putti* atop the Cavalcanti Altar by Donatello from 1435 are shown expressing fright at the height that they find themselves. Dempsey, *Inventing*, p. 38.

20   In several sketches the *putti* hold similar "ribbon-like bands" to those that the angels hold in the final solution.

21   Marder, *Bernini*, p. 42.

22   In the New Sacristy Tombs by Michelangelo a sketch depicted *putti*, but like much of the figural ornament, *putti* were not included in the built work.

23   In Michelangelo's sketches for the New Sacristy tombs, figures often appear to have a dynamic presence. *Putti* are depicted at the bottom of the broken pediment in an elevation design drawing (Paris Louvre 838) and the assertion here is that they arrived there by sliding down the pediment.

24   Pen-and-ink sketch now at Windsor Royal Library attributed to Bernini's workshop, and described by Lavin as "the earliest extant drawing for Urban's tomb." Irving Lavin, *Drawings by GianLorenzo Bernini* (Princeton, 1981), p. 62.

25   *Charity with two children*, by Gian Lorenzo Bernini. Terracotta. 41 cm. c.1634. Museo Sacro Musei Vaticani (Vatican City, Italy).

26   Trim marks in the clay indicate the possibility that the figures were sculpted in conjunction with an architectural surround; Putto_1435 refers to deformations and indentations that were caused by this process. The findings supporting the scenario of the making of the model in conjunction with the architectural surround are described in C.D. Dickerson and Anthony Sigel, "Charity with Four Children," in *Bernini: Sculpting in Clay*, ed. Nancy Grubb (New York, 2012), pp. 112–17.

27   The bronze bees represent the Barberini family, to which Pope Urban VIII belonged.

28   The skeleton is widely understood to represent death; it was introduced to the design in sketches following the pen-and-ink wash. A similar skeleton appears again in the Tomb of Pope Alexander the VII, and also in St. Peter's Basilica.

29   The Four Doctors of the Catholic Church and the Effigy of Pope Urbino VIII stand on either side of Putto_1435.

30   In this case, to call Bernini's design a frame is inadequate; "surround" refers to an entity that surrounds or frames.

31   "There is no relationship of matter and form between the angel and the body he assumes to communicate with men. The angel is like a motor for the body; his mobile body is only the outward representation of that motor." Giovanni Careri, *Bernini: Flights of Love, the Art of Devotion* (Chicago, 1995), p. 21.

32   Sketch referred to by Gian Lorenzo Bernini, *Study of a Putto in the Gloria above Cathedra Petri* (392 × 227 mm), c.1660. Black chalk on white paper. Liepzig 7900r.

33   The three phases of the masses expansion are recognized by Lavin, *Drawings by Gianlorenzo Bernini*.

34   Such preparatory drawings with perspective grids as the ground include Leonardo da Vinci's drawing for *Adoration of the Magi* of 1481.

35   Putto_1435 is referring to the mesh shell that defines the shape of his body in the animation software that helped to initially conceive him. Putto_1435

36   Indeed, Body Agent Aedicula is a non-denominational installation, where visitors are encouraged to "see themselves seeing."

37   Here Putto_1435 refers to a series of digital animations that were used to design the project.

38   Imprints of Putto_1435's body were left in the ornamental surface cladding of Body Agent Aedicula.

39   The reference to *putti* appearing to fly upwards in Bernini's designs is recognized by Giovanni Careri in the Fonseca Chapel where they appear to "advance upward." Careri, *Bernini*, p. 26.

40   The official name for the project that Putto_1435 refers to as the "Nuovo tempietto" is the Body Agent Aedicula.

41   The concept of habitus, coined by Marcel Mauss and elaborated by Pierre Bourdieu, refers to the predispositions of particular individuals or social groups acquired through the activities and experiences of everyday life and anchored in the body.

42   The constituent elements of the Baroque are described disparagingly by Lars Spuybroek as fixed forms that are "classical I-figures" and a material that is the "ground, finest white powder." Spuybroek, *Sympathy*, pp. 44 and 66. An alternative description of the Baroque conceptualization involving a tension between fixed forms and materiality is that ideal geometric forms were thought to be comprised of ethereal heavenly substances: Hersey, *Architecture*, pp. 18–21.

14

Communication Material: Experiments with German Culture in the 1930 Werkbund Exhibition

Sandra Karina Löschke

If there are architectural investigations that center on the 1930 Werkbund Exhibition, they tend to direct our attention to standardized designs of modern buildings and everyday objects that deployed the mass-produced dream materials of the time: various kinds of plastics, glass, and aluminum were celebrated as the achievements of an emerging high-industrial society. The exhibition's compelling economy of function and material conjured up a vision of rationalization, mass production, and unstoppable progress made possible by "the practical interpenetration of economy and technology with the world of art" that had been advanced by the Werkbund and the Bauhaus over the past decades.[1] "From stationery to chair, from cup to furniture covering, from theatre to sports field, and from the newest Hanomag locomotive to the serially-produced car," the exhibition represented outstanding examples of German design as part of a new, all-embracing techno-aesthetic system.[2]

Counter to the survey nature of the exhibition, there is, however, a second consideration that looks beyond the particularity and material logic of the individual designs. Instead, it considers materiality at the level of the exhibition, which, with its intense atmosphere and innovative display techniques, can be seen as an edifice in its own right. Lined up, stacked, and suspended mid air, the exhibition material generated patterns and rhythms across walls, ceilings, and rooms that structured the visitor's experience. Together with dynamic illumination, reflective materials, and colorful neon signs, the staging of material appears to have aimed at the production of architectural experiences that translated the concepts of modernity as sensory impressions.

The aim of this chapter is to pursue the ways in which the Werkbund Exhibition signified a shift away from the formal representation of objects and towards their use in the aesthetic staging of material sensations and experiences. The five exhibition spaces can be construed as a series of discrete perceptions, which constituted the actual building material of the exhibition—"blocs of sensation," to use Deleuze and Guattari's term.[3] In mapping the modernist logic of innovative materials onto the aesthetics of display, the curators made the audience believe in simultaneity between technological progress, social ideals, and modern life experiences. Focusing on the arrangement of exhibition spaces and their reception, this chapter argues that, in the diverse approaches of the curatorial team, the momentum of a convergence between material technology and material imagination was building up towards a display paradigm that significantly transfigured curatorial practice from representation to mediation.

THE WERKBUND EXHIBITION IN PARIS 1930—BACKGROUND

Germany's participation in the 1930 Werkbund Exhibition signaled a positive development in the re-establishment of Franco-German relations, which had been severely affected by the traumatic events of World War I. In its aftermath, France had not extended invitations to Germany for previous exhibitions and thus Germany's participation under the organizational leadership of Walter Gropius and the Werkbund elicited high expectations on all sides.[4] In his script for the exhibition, Gropius declared that the exhibition would not only present Germany's recent advancements in design but also convey an image of a new German mind-set:

> The German section at the "Exposition de la Société des Artistes Décorateurs" will
> be a small demonstrative show that will testify to this kind of mental attitude
> in Germany today; it will present methods and results that will register the
> development of direct connections to today's social and technical world in the
> areas of architecture, living, theatre and individual objects from furniture to
> jewellery.[5]

This "practical interpenetration of economy and technology with art," Gropius elaborated, had overcome the national and personal emphasis in art practices that had dominated the Wilhelmine era—a sentimental, decorative, and academically rooted tradition, which had now been replaced by the Werkbund's singular vision and innovative attitudes. The latter had not only experienced a dramatic rise in interest over the past decade but had indeed "captivated the entire country," he continued.[6] Thus Gropius presented the appointment of the Werkbund as the logical consequence of historic developments in Weimar Germany.

The spaces assigned to Gropius and his team were less than ideal for an exhibition. As Sigfried Giedion reported, the "Section allemande" was located at the far end of the neo-baroque Grand Palais with almost no opportunities for natural illumination.[7] Gropius provided the overall concept: a series of five seamlessly interconnected exhibition spaces, each thematizing a particular aspect of German cultural production. Whilst Gropius assumed responsibility for the first room, he assigned the planning of the other four rooms to his former Bauhaus colleagues, ensuring a consistent approach that represented the Bauhaus ethos under his directorship, 1919–1928. The first room showed a full-scale mock-up of the communal areas of a 10-storey residential building designed by Gropius for his "Genossenschaftsstadt" project. In the adjacent space (Room 3), Marcel Breuer demonstrated what a model apartment for a couple without children in an apartment hotel could look like.[8] Occupying the great double-height hall,

Gropius and Breuer's rooms formed the center of the exhibition from which the visitor entered the three side rooms organized by László Moholy-Nagy and Herbert Bayer. In Room 2, Moholy-Nagy presented developments in photography, lighting, and theater; and Rooms 4 and 5 were dedicated to serially produced goods and architectural projects arranged by Bayer. With a wide spectrum of themes and display strategies, ranging from the display of 1:1 mock-ups to the staging of small-scale objects, the five rooms were formally unified firstly, by a select palette of innovative materials, and secondly, by the imaginative staging of material to achieve maximum sensory impact.

THEORETICAL POSITIONS: MATERIAL LOGIC AND MATERIAL IMAGINATION

By the 1920s, materials had rapidly moved from the realm of craft to serial prefabrication: if previously manual and technical skills were required, now intellectual consideration was the key. Gropius compared designing with prefabricated standard elements to playing with "a building set at a large scale," which, he believed, offered the opportunity to combine perfectly fitting elements in endless variations.[9] Modern design depended on "spatial conception and the massing of its parts," as much as on "the limits of mechanics, statics, optics and acoustics," Gropius explained in a typescript for the Werkbund Exhibition.[10] This quasi-scientific outlook was also noted by critics who discerned that in the exhibition "the engineer-like is transformed into a serious play with form fantasy."[11] The result of this attitude was neither purely technical nor purely aesthetic, as French critic Paul Fierens remarked.[12]

In short, material logic and material imagination emerged as the driving forces behind the curatorial concepts of the exhibition—if in distinct ways: part of the exhibition pictured materials as the key to modern design, which, as Gropius proposed, was no longer based on the "external addition of decorative ornamentation and profile" but on the interrelations between material, mass, and color.[13] These interrelations constituted an object's unique material logic and provided a design concept that could be extended towards the "'design of life processes' and thus the uniform shaping of the entire spatial world on the basis of its social, aesthetic, technical, and economic preconditions," as Gropius stipulated in an exhibition text.[14] In his text, he further outlined a techno-aesthetic paradigm, which substantiated the modernist fiction generated by the Bauhaus that the material logic of technology also possessed its own aesthetic appeal. The Werkbund Exhibition, and in particular, his and Breuer's designs for the great hall (Rooms 1 and 3), were intended to deliver the empirical proof for this hypothesis— it was to be a "demonstrative show," as Gropius insisted.[15]

Other sections of the Werkbund Exhibition recognized the powerful mediating component of materials that existed independent of utilitarian considerations. If Gropius focused predominantly on the techno-aesthetic advantages of new materials, Moholy-Nagy imagined their mediating power by linking materials directly with sensory effects and their impact on human consciousness. In a manifesto titled "Dynamisch-konstruktives Kraftsysstem" (Dynamic–constructive Force-system),

Moholy-Nagy registered what he understood to be the underexploited mediating power of materials:

> we have to replace the static principle of classical art with the dynamic principle
> of universal life. Practically: instead of the static material-construction (material-
> and form-relations) the dynamic construction (vital constructivity, relations of
> forces) must be organized, where material is deployed as a mediator of energies
> only.[16]

For Moholy-Nagy, something beyond a purely technological aesthetic was at stake, something with much broader socio-cultural potential.[17] Dynamic elements such as spatial installations and film, he propositioned, could intensify perception and mental alertness and thus catapult the viewer out of a passive mode of reception.[18] Unlike Gropius, Moholy-Nagy was not interested in the material object itself but in its manifold effects. To probe the dynamic interrelations between audience, material, and space, Moholy-Nagy developed the light–space modulator—a light projection machine that transformed its environment by means of colorful light effects. Shown for the first time at the Werkbund Exhibition, the apparatus represented Moholy-Nagy's move towards an art that was quasi "non-material." The knowledge required for this medial intervention was both technical—to achieve these effects—and psychological—to direct and calculate their influence upon the viewer. Compared to the investigation of physical laws in engineering and science, Moholy-Nagy felt that "research into psycho-physiological laws of optical effectiveness [was] far behind" and had just begun to be explored, and he saw himself at the forefront of these endeavors.[19]

The psychological impact of exhibitions on the visitor also formed the focus of Herbert Bayer's work, which was informed by his robust engagement with graphics and advertising. For Bayer, the relationship between exhibition and visitor was regulated by the criteria of vision. Because "the exhibition space is available to the individual eye," Bayer determined that it "should obtain its forms from the qualities of the eye itself."[20] As a temporary space, the exhibition could focus on what was central for its success—the intensity of experience and its control:

> The theme should not retain its distance from the spectator, it should be
> brought close to him, penetrate and leave an impression on him, should explain,
> demonstrate, and even persuade and lead him to a planned and direct reaction.
> Therefore we may say that exhibition design runs parallel with the psychology of
> advertising.[21]

Whatever form an exhibition might take, whatever its theme, audience, or order, Bayer believed that it was never innocent or objective. More accurately, it was a venue where impressions and reactions were strategically planned and staged according to the psycho-technical laws of advertising.

This sense of an expanded understanding of materiality that was not object oriented but based on its mutual relation with the human body and psyche pervaded the majority of Bauhaus curatorial practice and design. Oskar Schlemmer's article "Gestaltung aus dem Material" (Design on the Basis of Material) presented the material conditions of the environment as the basis for corporeal experience: "Center, corners, partition, height, width, and depth, and emanating from this, the invisible net of lines ... Material! In the same sense material can be and can become: color, structure, facture of the floor surface, wall surface (timber, textile), and incidence and brightness of light etc."[22] For Schlemmer, the human assimilation of material surroundings allowed the development of an intensified "Raumgefühl" (feeling for space). Consequently, "a point in space, a surface, a colored surface, a plastic object, a shining moving object, a light space, a dark, low, divided, symmetric, asymmetric space, stairs, maps, furniture"—all were "injections into the body" and effected "the most diverse corporeal reactions," Schlemmer proposed.[23]

In summary, the re-imagination of materiality as relational and based upon psychological control, provided Moholy-Nagy, Bayer, and Schlemmer with the enabling assumptions for an expanded curatorial model that focused on the skillful mediation of images. What remained subject to debate, was how precisely effective mediation was to be achieved and how the interaction of the visitor with both the exhibited objects and the material environment as such could be stage-managed to exacting ends.

INNOVATIVE MATERIALS

Eager to be seen as affirmative of technological progress, Gropius and his colleagues demonstrated their preference for the most advanced industrial materials. With a view to his collaboration with the metal industry for the upcoming 1931 Deutsche Bauausstellung (German Building Exhibition), Gropius promoted non-ferrous metals as part of the Paris show.[24] These were used for the glass display elements, and various aluminum profiles from Wieland Werke in Ulm were displayed by Herbert Bayer in the section for serial products in Room 4. At the time, metal composites were frequently associated with notions of luxury and exclusivity: Nirosta, a new metal composite by German steel manufacturer Krupp Stahl, was promoted in its own lavish showrooms as a corrosion-free metal whose surface properties offered a new aesthetic appeal. Gropius did not fail to praise these benefits: their practical advantages—"their homogeneity, weather-resistance, water-repellence, and rust freeness"—found their aesthetic correspondence in attractive surface qualities; unlike iron, non-ferrous metals did not require a paint coating, and the "desired effects" could be achieved "with the natural color and surface of the material itself," he proposed.[25]

Advanced chemical processes led to the invention of plastics such as Trolit—a panel that could be produced with reflective, matt or varicolored surface qualities, and was intended to replace gypsum board and traditional wallpapers. An exhibition review by Max Osborn described Trolit as an "excellent material" that could "be produced in any size or form" and could assume the appearance of "giant tiles, or stucco lustro, or even real stone whilst all the same retaining its specific texture."[26] When left unglazed, he continued, Trolit displayed "an epidermis much like biscuit porcelain." Whilst most of the tiles he saw in the exhibition were of a "milky coloration," he was impressed with a small number of sheets in "strong colors" that calmly floated like "airy ornaments" within the pale walls. Trolit featured across the entire exhibition in variety of scenarios—Gropius's cost plan listed 150m² of Trolit panels for the communal areas alone.[27] Moholy-Nagy specified it for a thin, two-sided screen enclosing the "Kino-Box" (cinema box), where the panels appeared in a patchwork of colors on the external side, their high gloss finish generating an almost fluid appearance (Figure 14.3).[28] On the inside of the box, the finish was more restrained and the choice of a matt black surface avoided reflections and created a darker environment for the slide projections.

Progress in glass manufacture brought forth diverse types of glazing and increased sheet sizes with ever-slimmer frames lent maximum transparency to display windows and buildings. An article published in *Die Baugilde* credited the glass display cases at the Werkbund Exhibition for pushing the boundaries of what seemed technically conceivable in the field of glass technology: "The free-standing glass display case, [that is] the show box which is viewable from all sides and made from metal and glass with a maximum of transparency, had to become even more transparent," they reported.[29] The cases featured a metal three-corner profile designed to receive a beveled glass edge—a patented detail that achieved a completely flush transition between metal and glass, and created a smooth external skin that appeared to wrap near seamlessly around the display (Figure 14.1).

14.1   Corner detail of glass display case— Werkbund Exhibition, Room 2.

2 Eckschnitt einer Vitrine mit Dreikantprofil

3 Schnitt durch die Tür

1 Vitrine in der Werkbund-Ausstellung, Paris (Prof. Dr. Walter Gropius, Architekt BDA, Berlin). Ausführung: Alex. Herman G. m. b. H., Berlin N 20

14.2   View of curved glass screen and projector behind the "Kino-Box" in Room 2.
Laszlo Moholy-Nagy, design.

The mandate of technological innovation, which exhibitions carried, is explained by Bayer:

> *In addition, different materials require different methods of handling and constructive and special effects. Here, too, important experience was gained in the practical construction of commercial buildings and exhibitions … Entirely new techniques were discovered and investigated. In this field of exhibitions, the fruits of universal education were apparent: familiarity with many different techniques and materials and their possibilities in design.*[30]

Whereas industry conceived the minimization of material as an attempt not to "distract the visitor with details" and instead "focus his gaze on what is offered within in the glass- and metal-enclosed space," the photographs of the exhibition testify to the opposite.[31] In their interplay with artificial light, reflections etc., the glass display cases were the actual stars of the exhibition and appeared as a number of different types—with solid sheet infill, colorful bases, variances in steel framing, size, and height. Bayer's enormous glass display for serial products, for example, outsized and upstaged the small objects it contained, tipping the balance between exhibit and exhibition architecture in favor of the latter.

In the second room, Moholy-Nagy installed a curved glass screen made with profiles from the Wieland Werke. This strategically placed dividing screen amplified the brightness produced by the mass display of lighting fixtures along the opposite wall, as well as the colorful light play of the light–space modulator and the "on–off" effects of the automatic slide projections in the "Kino-Box" (Figure 14.2). Not all experiments with glazing attained the desired effect and some were dismissed as "entertaining" but "not to be taken seriously"—as was the case for a matt glass sheet with concave lenses that showed the space on the other side "in little pictures," as Parisian architect and critic Durand-Dupont somewhat sarcastically remarked.[32]

STAGING THE FUTURE: GROPIUS AND BREUER'S PROTOTYPES FOR MODERN LIVING

If the play with modern materials and technologies unfurled real, and at times patentable solutions, the other "material" of the exhibition was perceptual in nature and produced spectacular images rather than showcasing technological or practical outcomes. One of these material showcases was a ramp and elevated walkway made of galvanized Tezett metal grids. Sigfried Giedion described it as part of his first impression upon entering the exhibition: attention was immediately directed to a swimming pool, above which rose a stair and elevated walkway made of Tezett metal grids, leading the visitor up and on to subsequent exhibition rooms.[33] For Wilhelm Lotz, the Tezett construction was a "material demonstration." However, he questioned whether the material was really a practical choice for this purpose. At a combined cost of 24,000 Reichsmark, stair, ramp, and elevated walkway were certainly the most expensive exhibition items.[34]

If the novelty value of the material caused great astonishment, this effect was upstaged by the provision of unusual experiential dimensions: filtered views through the metal grid floor from above and below caught the attention of the visitors—if not always for the right reasons, as Giedion observed.[35] The high visual transparency of the material stimulated the visitor both physically as well as visually: walking across the open Tezett grid would have increased the visitor's awareness of height, and downward or upward views through the grating would have revealed almost photographic images. The Tezett flooring, it seems, reconstructed spatially the optics of contemporary photographic experimentation that used visuals grids and unusual perspectives from below or above—features that were distinctive for the work of Moholy-Nagy, Man Ray, and others, and which can be seen in the photographic documentation of the exhibition itself.[36]

Apart from its materiality, the elevated Tezett construction aided the formation of visual impressions. From the bird's-eye perspective a panoramic overview of the exhibition spaces unfolded in front of the visitor. Beholding the exhibition at a distance eliminated all detail and the onslaught of distracting sensory impressions to which the visitor was exposed on the ground. The change in elevation revealed the rhythmic structure of display elements, color distribution, and other visual phenomena that could not be detected from the ground and allowed the viewer to distil a pure image of the exhibition spaces—an abstraction of geometries, color, and light.

If elevated viewing underscored the conceptual aspects of the exhibition, the arrangement of materials at ground-floor level was directed at immediacy and intensity of impression. To provide an adequate setting for the full-scale prototypes in Rooms 1 and 3, Gropius had ordered large quantities of muslin fabric and linoleum to screen existing walls and the original mosaic floors, thus creating a neutral background for the actual displays.[37] With the background set, a variety of feature materials were ordered. Gropius's cost plan for Room 1 features crystal glass, mirrors, rippled glass sheets for the pool basin, and *luxfer* prism ceilings, amongst others.[38] In distinction to the monochromatic photographs of the exhibition, the material list testifies to the richness of the materials and colors. If the aesthetic and novelty of new industrial materials figured prominently in the media coverage of the exhibition, it was their application in the context of domestic environments that constituted the main cause for astonishment. The shock value of applying the same materials across previously distinct domains resided in the conflation of the home with typologies belonging to the non-private sphere (commercial, industrial etc.). Indeed, Gropius and Breuer's apartment spaces could not have been more different from regular domestic spaces at the time. "No more wall paper, dust-collecting curtains, and heavy, space-consuming timber furniture," Hans Heilmaier announced; instead the visitors found "steel, glass, Trolit, and linoleum," materials that in his mind addressed all criteria for modern living—"durable, rational, hygienic, and affordable."[39] Whilst the industrial aesthetic was deemed appealing by some reviewers, it was rejected by others: *La Liberté* (May 20, 1930) found the materials "horriblement dépourvu de charme" (horribly depraved of charm); *Daily Mail Paris* (May 2, 1930) conceded that they might be "hygienic but not beautiful"; and *L'Opinion* (June 7, 1930) believed they represented "la sécheresse de la machine" (the dryness of the machine).[40]

If the futuristic palette of materials elicited a range of reactions, the opportunity for testing these environments was received with general enthusiasm. In the full-scale exhibits, movable desks, bouncy cantilevered chairs, and other technical gadgets "gave the audience the opportunity to engage with this form of living as a whole" and this "happened not without appreciation," as Sigfried Giedion noted.[41] The de facto testing of the apartment prototypes thus not only offered immediate experience with innovative materials and technologies, but beyond that, it communicated the impression that one had indeed encountered the future—"time here becomes space," as Dr. Gutmann recapitulated in his review.[42]

THE CAPTURE OF THE EYE: BAYER'S MATERIALIZATION OF THE SPECTACULAR

In distinction to Gropius's and Breuer's pragmatic approach, Bayer's display strategy drew on visual communication tactics to realize what he regarded as the central challenge of the exhibition—"making non-visual ideas visible."[43] In a bid to overcome the "never-ending attack of influences, messages, and impressions" assailing the visitors in the modern world, the displays were to correspond to "a desired sequence of impressions" that was matched to "the visitor's abilities of perception," Bayer advised.[44] Based on a convergence of advertising and artistic practices, his curatorial ethos proposed a complex language of interrelated spatial and visual criteria:

> … language as visible printing or as sound, pictures as symbols, paintings, and photographs, sculptural media, materials and surfaces, color, light, movement (of the display as well as of the visitor), film, diagrams, and charts. The total application of all plastic and psychological means (more than anything else) makes exhibition design an intensified and new language.[45]

Apart from combinations between individual "linguistic" elements, Bayer foresaw a high degree of synthesis between them, whereby graphics became architectural, advertising psychology spatial, and color dynamic. Having received training as an architect, Bayer recognized that exhibition design offered opportunities to transfigure the primarily visual encounters of painting, graphics, photography, and advertising into spatial experiences—"in exhibition design," he had come to see "a new and complex means of communication," in which "all other known means of design" could be strategically recast to coagulate as the "apex of all collective effects, of all powers of design."[46] In bundling multiple media and fields of reference to converge upon the exhibition theme, Bayer endeavored to attain a maximum of intensity. A particularly relevant example is his display of serially produced chairs in Room 5 of the Werkbund Exhibition (Figure 14.4). By mounting chairs vertically across the wall, Bayer evocatively positioned them according to the modernist logic of industrial production and standardization promoted by the Werkbund—as a mass display, and arranged in rows.[47] Bayer's chair exhibit highlights techniques that were central elements of his display rationale—"the material,""the perspective of the individual," and "the movement of the individual."[48] Removed from its habitual association with the floor and re-associated with the visual plane, the chairs were seen, quite unnaturally, at eye level or from below. Showing them from atypical perspectives destabilized their familiar appearance, whilst their repetition in rows transfigured them into abstract patterns, visual structure, and material effect.[49] Bayer, it seems, consciously considered materiality as a display technique, suggesting that its effects extended above and beyond the physical material of the objects themselves and possessed the "same psychological and physiological functions as color." Under the logic of perception, Bayer believed wall, color, lighting, and chairs coalesced as a singular material consideration that communicated the concepts of modernity in its own compelling ways—"a psychology of effect" constituted upon a "discipline of the feeling for material" that, he believed, was "especially fostered at the Bauhaus."[50] Entertaining aesthetic conceits that expanded the modern logic of design and production towards new ways of seeing and experiencing, Bayer clearly felt that his displays could not simply be understood as a representation of modern product design and materials but that they provided a glimpse of the infinite possibilities of dynamic perception, attuning the visitor to see the world anew.

The arresting nature of his exhibit was widely noted. Sigfried Giedion felt that the arrangement stood out: the chairs, he noted, appeared to be "semi-levitating." This "capture of the eye," as Giedion termed it, was repeated "in the arrangement of the beautiful enlargements of new buildings, by arranging the images in a curvilinear fashion and at an angle that corresponded to the spatial perception of the eye."[51] Like the chairs, the architectural images and the serial objects arranged by Bayer, were equally attuned to the visitor's perceptual capacity. For Giedion, these displays exemplified Bayer's exceptional skill in the "suggestive arrangement of serially-produced objects."[52] What appears to be suggested by Bayer's strategy is the presentation of everyday objects not as art, but as seen with the material sensitivity of the modern eye.

MULTI-MEDIAL DISTRACTIONS: MOHOLY-NAGY'S CONCEPT OF MATERIALITY AS MEDIUM

Moholy-Nagy's 1922 text "Dynamic–constructive force-system" had formulated the parameters for his practical explorations of the interrelations between material, viewer, and space, which he proceeded to test in a variety of environments with custom-made apparatus. The most famous of these is the light–space modulator, originally called "Lichtrequisit einer elektrischen Bühne" (Light Requisite for an Electrical Stage) and shown in the second room of the Werkbund Exhibition. The apparatus drew on the latest in lighting technology and had been developed in collaboration with AEG (Allgemeine Elektrizitäts-Gesellschaft)'s theater department since 1922. It contained three moving platforms rotating in different speeds to each other and was constructed with the same modern materials that were used throughout the Werkbund Exhibition: transparent materials (zelon, matt and clear glass), materials pervious to light (transparent gauze, wire netting), and perforated materials (perforated nickel-plated brass plates). Yet in distinction to other exhibition displays, the materials of the light–space modulator were not deployed for their aesthetic appearance but for their capacity to mediate between viewer and environment—they reflected, deflected, and transmitted light and color in a dynamic manner, completely transforming the surrounding space. In contemporaneous reviews, the machine was not described in any detail—Grünberg made reference to it in his article "Die Gropius Ausstellung in Paris," which appeared in *Berliner Tageblatt* (May 31, 1930); and Hans Heilmaier briefly discussed it in "Deutsche Raumkunst triumphiert in Paris," published in *Neue Pariser Zeitung* (May 24, 1930). Because of the lack of commentary, it has been speculated that the modulator might have been hidden behind a fabric screen.[53] Yet descriptions by visitors testify to its perceived impact: a review by Dr. Gutmann recalled the lasting impression that the exhibition's colorful light made upon the visitor, admiring "the precision of forms, and the white, denuding light," and noting that, "when there are colors—strong, and luminous—they are not harmonically resolved and relaxing, but, with joyful force, alive and enlivening."[54] Max Osborn's review vividly described the intense sensory stimulation he experienced in Room 2: "Dynamics are offered instead of rigidity … Theatre models … rotate, flicker disquietingly through little lights … it glints and sparkles all around."[55] In the almost windowless spaces of the Werkbund Exhibition, Moholy-Nagy recognized the opportunity for elevating light as a key element for intensifying the visitor's experience—both as an essential requisite for illuminating the displays and as an instrument for creating transformative spatial effects that integrated the visitor as part of the work.

Yet, the possibilities for the light modulator's applications outside the theater were not acknowledged by any of the reviewers, and it was left to Moholy-Nagy to explain the full range of anticipated uses in a subsequent article, which he wrote for the Werkbund publication *Die Form* shortly after the exhibition.[56] In the introductory paragraph, Moholy-Nagy called attention to the design possibilities opened up by electricity: "Adjustable, artificial, electric light permits us to create rich light effects without much effort," he explained. These are not random effects: "with electric energy pre-calculated movements can be performed, that can be repeated precisely again and again."[57] Light and movement were becoming elements of design again, Moholy-Nagy predicted, but now considered in relation to the present situation: the festivities of the Baroque with their water fountains and stage sets could be re-invented in a contemporary context as light fountains and dynamic mechanic–electric plays, he envisaged. Moholy-Nagy predicted a variety of applications for his electro-mechanic spectacles: as advertising, as popular entertainment at fairs, or as a means to create moments of suspension in the theater. However, he also expanded the use of dynamic lighting from collective experience of public venues to the privacy of everyday life: in domestic applications, light requisites could be remote-controlled via radio, he imagined, or regulated by the owners with a set of exchangeable cardboard templates. These would be received as supplements in newspapers and provide patterns for dynamic light plays in yellow, green, blue, white.[58] The wider implications of Moholy-Nagy's ideas appear not to have been fully understood by visitors and critics alike, who saw the requisite mainly as a light machine for the theater. Yet in light of the desolation of many urban areas and poor living conditions at the time, which Moholy-Nagy later captured on film, it is possible to see their potential as affordable and instant

14.3   Werkbund Exhibition Paris 1930, view of Room 2 through the glass wall (Trolit screen of "Kino-Box" on the left). Laszlo Moholy-Nagy, design.

14.4   Werkbund Exhibition Paris 1930, Room 5: Exhibition of modern, standardized chairs. Herbert Bayer, design.

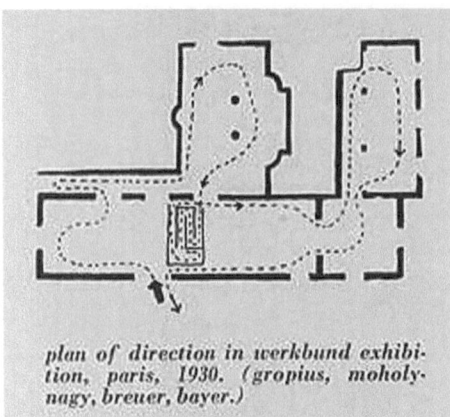

plan of direction in werkbund exhibi-
tion, paris, 1930. (gropius, moholy-
nagy, breuer, bayer.)

14.5 Diagram
of itinerary/
walking direction
in Werkbund
Exhibition.

transformations of space for collective spectacles or private enjoyment.[59] Oliver A.I. Botar described the modulator's application in a domestic environment as "a kind of disco ball for the home."[60] Whilst this was one of its possible uses, for Moholy-Nagy, the significance of the apparatus resided in its capacity to be developed towards a system.[61] Testing its potential for different applications, Moholy-Nagy variously adapted the machine for the transformation of spaces, the generation of photographs, and the production of films.[62] As he had already announced in his 1922 manifesto, "the development of the dynamic individual construct results in the DYNAMIC–CONSTRUCTIV FORCE SYSTEM." The system replaced the artwork, he continued, and the viewer "who was receptive in the beholding of previous artworks, now with all his capacities amplified, becomes himself an active factor in the unfolding of forces."[63] The modulator's capacity for the systematic aestheticization of everyday reality aimed at the erosion of boundaries between art and life towards a total work of art. Not surprisingly, it was Sigfried Giedion who, more than a decade later, proposed similar interventions as modern "democratic monuments."[64]

A NEW IMAGE OF GERMAN CULTURE AND GERMAN PEOPLE

Beyond addressing a wider desire for optical spectacles, the Werkbund Exhibition's imagery operated at an ethic–political register. Clean, hygienic, gleaming, futuristic, it presented a new material sensibility as proof of an altered German national conscious. Thus, as suggested by Gropius in his writings, and as frequently repeated by the press, the show delivered an externalized image of the collective psyche of the German people and their nascent democracy. French critics, such as Grünberg, reported to have sensed a "strong mental energy" emanating from the exhibition. To support his argument, Grünberg quoted from a review that had appeared in *Temps*:

> It is no exaggeration to say that the German section of the Grand Palais truthfully mirrors the face of the new Germany ... The interest in this exhibition is so great because it does not only give us an understanding of the new German aesthetic, but also insights into the collective psychology of the Germany of today.[65]

Connections between national identity and curatorial strategy were highlighted also by Siegfried Giedion. The powerful imagery of the exhibition, Giedion observed, went "beyond the sensory appeal of the objects on display" by imparting the idea of a "collective spiritual attitude" to the visitors.[66] In other words, it was not the aesthetic or practical value of the exhibits themselves, but Gropius and his co-curators' skill in systematically staging these. Giedion appears to point to something that resides in the selection and staging of material and that is difficult to adequately determine. This excess value of the exhibition can be described as a matter of material aesthetics and equates to the effective self-presentation of Weimar culture: in the imagination of the audience, the Werkbund Exhibition successfully positioned aesthetics and politics, art and reality as interchangeable and identical.[67] As Giedion would observe years later, Gropius and his collaborators recognized that "the architect had a part in forming the spirit of his times," and this required his engagement on a much bigger stage.[68]

BIBLIOGRAPHY

Bayer, Herbert, "Aspects of Design of Exhibitions and Museums," *Curator: The Museum Journal* 4, no. 3 (July 1961): pp. 257–88.

Bayer, Herbert, "Fundamentals of Exhibition Design," *PM (Production Manager)* 6, no. 2 (December 1939–January 1940): pp. 17–25.

Botar, Oliver A.I., "Gesamtkunstwerk ohne Kunst," in *Kunst Des Lichts: László Moholy-Nagy*, eds. La Fabrica (Berlin: Hirmer Verlag, 2010), pp. 159–68.

Böhme, Gernot, *Atmosphäre. Essays Zur Neuen* Ästhetik (Frankfurt am Main Suhrkamp, 2013).

Brunnhammer, Yvonne and Suzanne Tise, *Decorative Arts of France 1900–1942* (New York, NY: Random House, 1990).

Deleuze, Gilles and Félix Guattari, *What Is Philosophy?* (London: Verso, 1994).

"Die Deutsche Werkbundausstellung in Paris," *Metallwirtschaft* 9, no. 28 (July 1930): p. 599.

Dorner, Alexander, "Gedanken Zur Französisch-Deutschen Ausstellung in Paris," *Hannoverscher Anzeiger* (July 6, 1930).

Driller, Joachim, "Bauhäusler zwischen Berlin und Paris: Zur Planung und Einrichtung der 'Section Allemande' in der Ausstellung der Société des Artistes décorateurs Français 1930," in *Das Bauhaus und Frankreich, Le Bauhaus et la France, 1919–1940*, eds. I. Ewig, T.W. Gaehtgens, and M. Noell (Berlin: Akademie Verlag, 2002), pp. 255–74.

Durand-Dupont (alias Roger Ginsburger), "Der Deutsche Werkbund im Salon der 'Artistes–Décorateurs', Paris," *Das Werk*, no. 7 (July 1930): pp. 197–203.

Fierens, Paul, "Le 'Deutscher Werkbund' au Salon des Artistes Décorateurs," *Le Journal des Débats* (June 10, 1930).

Gebert, Jakob and Kai-Uwe Hemken, "Der Raum der Gegenwart: Die Ordnung von Apparaten und Exponaten," in *Kunst Licht Spiele: Lichtästhetik der klassischen Avantgarde*, eds. U. Gartner, Kai-Uwe Hemken, and Kai Uwe Schierz (Bielefeld: Kerber, 2009), pp. 138–55.

Giedion, Sigfried, "Der Deutsche Werkbund in Paris," *Der Cicerone* 22, no. 15/16 (August 1930): pp. 429–34 (illustrated republication of article first published in *Neue Zürcher Zeitung* (June 17, 1930)).

Giedion, Sigfried, *Space, Time and Architecture: The Growth of a New Tradition* (Cambridge, MA: Harvard University Press, 1967).

Giedion, Sigfried, José Luis Sert, and Fernand Léger, "Nine Points on Monumentality," written in 1943. Published in *Architecture, You, and Me: The Diary of a Development*, edited and written by Sigfried Giedion (Cambridge, MA: Harvard University Press, 1958), pp. 48–51.

Goergen, Jeanpaul, "Filme, Projekte, Vorschläge: Annotierte Filmografie 1921–1934," in *Kunst Des Lichts: László Moholy-Nagy*, eds. La Fabrica (Berlin: Hirmer Verlag, 2010), pp. 243–8.

Gropius, Walter, "Architektur auf der Deutschen Werkbundausstellung in Paris, Grand Palais, Mai bis Juli," typescript, 1 August 1930, Gropius Walter (1889–1969), Werkmanuskripte, GS20: Aufsatz- und Vortragsmanuskripte, Mappe 62, Bauhaus-Archiv Berlin.

Gropius, Walter, "Die Deutsche Abteilung auf der 'Exposition de la Société des Artistes décorateurs' in Paris, Grand Palais, im Mai 1930," typescript, 1930, Gropius Walter (1889–1969), Werkmanuscripte, GS20: Aufsatz- und Vortragsmanuskripte, Mappe 63, Bauhaus-Archiv Berlin.

Gropius, Walter, "Kostenzusammenstellung für den Raum I der Ausstellung Deutscher Werkbund Paris," unpaginated typescript, 1930, Gropius Walter (1883–1969), GN Kiste 6, Mappe 318: Deutscher Werkbund 1930, 79. Exhibitions, Bauhaus-Archiv Berlin.

Gropius, Walter, "Nichteisen-Metalle in der Bauwirtschaft," typescript, October 1930, Gropius Walter (1889–1969), Werkmanuskripte, GS20: Aufsatz- und Vortragsmanuskripte, Mappe 69, Bauhaus-Archiv Berlin.

Grünberg, J., "Die Gropius Ausstellung in Paris," *Berliner Tageblatt* (May 31, 1930).

Gutmann, Dr. U., "Zum Raum wird hier die Zeit (Zur Eröffnung der Ausstellung des Deuschen Werkbundes in Paris)," *Königsberger Hartungsche Zeitung* (May 20, 1930).

Heilmaier, Hans, "Deutsche Raumkunst triumphiert in Paris," *Neue Pariser Zeitung* (May 24, 1930). Press clipping from Findbuch Harvard College Library, Walter Gropius, Papers II, Press Clippings 1930, 29/1–18. Ausstellung Deutscher Werkbund, Paris, 1930 (copy at Bauhaus-Archiv Berlin).

Lotz, Wilhelm, "Ausstellung des Deutschen Werkbundes in Paris," *Die Form: Zeitschrift für gestaltende Arbeit* 5, no. 11–12 (1930): pp. 281–96.

Margolin, Victor, *The Struggle for Utopia: Rodchenko, Lissitzky, Moholy-Nagy 1917–1946* (Chicago, IL: University of Chicago Press, 1997).

Merleau-Ponty, Maurice, *Phenomenology of Perception*, translated by Colin Smith (London: Routledge & Kegan Paul, 1962).

Moholy, Lucia, *Marginalien zu Moholy-Nagy* (Krefeld: Schrepe Verlag, 1972).

Moholy-Nagy, László, "Die beispiellose Fotografie," *Das Deutsche Lichtbild* (Jahresschau 1927): pp. 10–11.

Moholy-Nagy, László, "Die Photographie in der Reklame," *Photographische Korrespondenz Zeitschrift für wissenschaftliche und angewandte Photographie und die gesamte Reproduktionstechnik* 63, no.9 (September 1, 1927): pp. 257–60.

Moholy-Nagy, László, "Lichtrequisit einer elektrischen Bühne," *Die Form: Zeitschrift für gestaltende Arbeit* 5, no. 11–12 (1930): pp. 297–8.

Moholy-Nagy, László and Alfred Kemény, "Dynamisch-Konstruktives Kraftsysstem," unpaginated typescript, June–July 1922, Moholy-Nagy, Mappe 1, Werkmanuskripte, Bauhaus-Archiv Berlin.

Moholy-Nagy, Sibyl, *Moholy-Nagy: Experiment in Totality* (New York, NY: Harper & Brothers, 1950).

"Neue Formen in Metall und Glass," *Die Baugilde* (September 30, 1930): p. 1620. Press clipping from Findbuch Harvard College Library, Walter Gropius, Papers II, Press Clippings 1930, 29/1–18, Ausstellung Deutscher Werkbund, Paris, 1930 (copy at Bauhaus-Archiv Berlin).

Osborn, Max, "Gropius und die Seinen – Die Deutsche Ausstellung in Paris," *Vossische Zeitung* (May 22, 1930).

Overy, Paul, "Visions of the Future and the Immediate Past: The Werkbund Exhibition, Paris 1930," *Journal of Design History* 17, no. 4 (2004): pp. 337–57.

Schlemmer, Oskar, "Gestaltung aus dem Material," *Das neue Frankfurt: Internationale Monatsschrift für die Probleme kultureller Neugestaltung* 4, no. 10 (Ausstellungen, October 1930): pp. 222–5.

Staniszewski, Mary Anne, *The Power of Display: A History of Exhibition Installations at the Museum of Modern Art* (Cambridge, MA: MIT Press 2001).

Torp, Claudius, *Konsum und Politik in der Weimarer Republik* (Göttingen: Vandenhoek & Ruprecht, 2011).

NOTES

1 Walter Gropius, "Architektur auf der Deutschen Werkbundausstellung in Paris, Grand Palais, Mai bis Juli 1930," typescript, 1 August 1930, Gropius Walter (1889–1969), Werkmanuskripte, GS20: Aufsatz- und Vortragsmanuskripte, Mappe 62, Bauhaus-Archiv Berlin, p. 1.

2 Alexander Dorner, "Gedanken zur Französisch-Deutschen Ausstellung in Paris," *Hannoverscher Anzeiger* (July 6, 1930).

3 Gilles Deleuze and Félix Guattari, *What Is Philosophy?* (London, 1994), p. 164.

4 Paul Overy elucidated differences in reception by German and French media. See Paul Overy, "Visions of the Future and the Immediate Past: The Werkbund Exhibition, Paris 1930," *Journal of Design History* 17/4 (2004): pp. 337–57. Yvonne Brunnhammer highlights differences between French and German attitudes towards individuality and tradition in Yvonne Brunnhammer and Suzanne Tise, *Decorative Arts of France 1900–1942* (New York, 1990), p. 179.

5 Walter Gropius, "Die Deutsche Abteilung auf der 'Exposition de la Société des Artistes décorateurs' in Paris, Grand Palais, im Mai 1930," typescript, 1930, Gropius Walter (1889–1969), Werkmanuscripte, GS20: Aufsatz- und Vortragsmanuskripte, Mappe 63, Bauhaus-Archiv Berlin, p. 3.

6 Gropius, "Deutsche Abteilung," p. 3.

7 Sigfried Giedion, "Der Deutsche Werkbund in Paris," *Der Cicerone* 22/15–16 (August 1930): p. 429.

8 Gropius's model of modern living reflected recent societal changes that had eroded family living in favor of apartment living, shown amongst others in the increased number of single women who had joined the workforce after WWI and the small core family. It is comparable to ideas discussed in Le Corbusier's *Vers une Architecture*. Cf. Claudius Torp, *Konsum und Politik in der Weimarer Republik* (Göttingen, 2011), p. 210.

9 Walter Gropius, "Nichteisen-Metalle in der Bauwirtschaft," typescript, October 1930, Gropius Walter (1889–1969), Werkmanuskripte, GS20: Aufsatz- und Vortragsmanuskripte, Mappe 69, Bauhaus-Archiv Berlin, p. 1.

10 Gropius, "Architektur auf der Deutschen Werkbundausstellung," p. 3.

11 Max Osborn, "Gropius und die Seinen—Die Deutsche Ausstellung in Paris," *Vossische Zeitung*, May 22, 1930.

12 Paul Fierens, "Le 'Deutscher Werkbund' au Salon des Artistes décorateurs," *Le Journal des Débats* (June 10, 1930), cited as a German translation in Gropius, "Architektur auf der Deutschen Werkbundausstellung," p. 7.

13   Walter Gropius cited in Wilhelm Lotz, "Ausstellung des Deutschen Werkbundes in Paris," *Die Form: Zeitschrift für gestaltende Arbeit* 5/11–12 (1930): p. 281.

14   Gropius, "Deutsche Abteilung," p. 3.

15   Gropius, "Architektur auf der Deutschen Werkbundausstellung," p. 2.

16   László Moholy-Nagy and Alfred Kemény, "Dynamisch-Konstruktives Kraftsysstem," typescript, June–July 1922, Moholy-Nagy, Mappe 1, Werkmanuskripte, Bauhaus-Archiv Berlin.

17   On the social values in Moholy-Nagy's work and its reception in Weimar Germany, see Victor Margolin's book chapter on "The Politics of Form: Rodchenko and Moholy-Nagy, 1922–29," in *The Struggle for Utopia: Rodchenko, Lissitzky, Moholy-Nagy 1917–1946* (Chicago, 1997), pp. 123–62.

18   The transfer of dynamic elements from film to other areas of design was paradigmatic for Moholy-Nagy's design ethos, which was based on absolute intermediality. Recounting Moholy-Nagy's commission to design a window display for Simpson's Piccadilly in 1936, Sibyl Moholy-Nagy noted that for Moholy-Nagy "design was indivisible." Whether commercial window displays or stage sets for operas, for him, both entailed the same task: "they had to appeal to perception and emotion in the onlooker, just as do painting and sculpture. The message was different, but the sense apparatus to absorb it remained the same. Design was indivisible." Sibyl Moholy-Nagy, *Moholy-Nagy: Experiment in Totality* (New York, 1950), p. 125.

19   László Moholy-Nagy, "Die Photographie in der Reklame," *Photographische Korrespondenz Zeitschrift für wissenschaftliche und angewandte Photographie und die gesamte Reproduktionstechnik* 63/9 (September 1, 1927): p. 257.

20   Herbert Bayer, "Fundamentals of Exhibition Design," *PM (Production Manager)* 6/2 (December 1939–January 1940): pp. 23–4.

21   Bayer, "Fundamentals of Exhibition Design," p. 17.

22   Oskar Schlemmer, "Gestaltung aus dem Material," *Das neue Frankfurt: Internationale Monatsschrift für die Probleme kultureller Neugestaltung* 4/10 (Ausstellungen, October 1930): p. 223.

23   Ibid.

24   Anonymous author, "Die Deutsche Werkbundausstellung in Paris," *Metallwirtschaft* 9/28 (July 1930): p. 599.

25   Gropius, "Nichteisen-Metalle in der Bauwirtschaft," p. 3.

26   Max Osborn, "Gropius und die Seinen."

27   Walter Gropius, "Kostenzusammenstellung für den Raum I der Ausstellung Deutscher Werkbund Paris," typescript, 1930, Gropius Walter (1883–1969) GN Kiste 6, Mappe 318: Deutscher Werkbund 1930, p. 79. Exhibitions, Bauhaus-Archiv Berlin.

28   Giedion described the screen as "three walls made of a new kind of material (Trolit)," acting as "light protection" for the slide projections. Giedion, "Der Deutsche Werkbund in Paris," p. 431.

29   Anonymous author, "Neue Formen in Metall und Glass," *Die Baugilde* (September 30, 1930): p. 1620. Press clipping from Findbuch Harvard College Library, Walter Gropius, Papers II, Press Clippings 1930, 29/1–18, Ausstellung Deutscher Werkbund, Paris, 1930 (copy at Bauhaus-Archiv Berlin).

30   Bayer, "Fundamentals of Exhibition Design," p. 24.

31   Anonymous author, "Neue Formen in Metall und Glass," p. 1620.

32   Durand-Dupont (alias Roger Ginsburger), "Der Deutsche Werkbund im Salon der 'Artistes–Décorateurs', Paris," *Das Werk*, no. 7 (July 1930): p. 198.

33   Giedion, "Der Deutsche Werkbund in Paris," p. 430.

34   The cost for the Tezett construction is listed in Gropius, "Kostenzusammenstellung für den Raum I der Ausstellung Deutscher Werkbund Paris." With the average blue-/white-collar salary ranging between 40 and –45 Reichsmark per week in 1930, Tezett clearly was not an affordable building material at the time. The 1930 Weimar Republic salary figures are available at http://hsr-trans.zhsf.uni-koeln.de/hsrretro/docs/artikel/hsr/hsr1981_32.pdf [accessed 01/11/2010].

35   Giedion noted that women in particular felt uncomfortable being subjected to the gaze of other visitors standing below. Cf. Giedion, "Der Deutsche Werkbund in Paris," p. 430.

36   For Moholy-Nagy, photography focused on an entirely "new way of seeing" that worked with optical effects such as "unusual perspectives [achieved] through angled positions, up- and down-photography" and the "structure (faktura) of different materials," as he suggested in "Die beispiellose Fotografie," *Das Deutsche Lichtbild* (Jahresschau, 1927): p. 10.

37   In a protocol written during a visit to the Grand Palais, Marcel Breuer reported that the existing mosaic floors were not to be damaged. Exhibits, he suggested, should be fixed to walls, placed on the floor, or fastened to metal openings in the floor. See Joachim Driller, "Bauhäusler zwischen Berlin und Paris: Zur Planung und Einrichtung der 'Section Allemande' in der Ausstellung der Société des Artistes décorateurs Français 1930," in *Das Bauhaus und Frankreich, Le Bauhaus et la France, 1919–1940*, eds. Isabelle Ewig, Thomas W. Gaehtgens, and Mathias Noell (Berlin, 2002), p. 264.

38   Gropius, "Kostenzusammenstellung für den Raum I der Ausstellung Deutscher Werkbund Paris."

39   Hans Heilmaier, "Deutsche Raumkunst triumphiert in Paris," *Neue Pariser Zeitung* (May 24, 1930). Press clipping from Findbuch Harvard College Library, Walter Gropius, Papers II, Press Clippings 1930, 29/1–18. Ausstellung Deutscher Werkbund, Paris, 1930 (copy at Bauhaus-Archiv Berlin).

40   Newspaper excerpts from Walter Gropius's collection of press commentaries. See "Pressestimmen zur Deutschen Werkbundausstellung Mai–Juli 1930 im Grand Palais in Paris," typescript, undated, Gropius Walter (1889–1969), Bauhaus-Archiv Berlin.

41   Giedion, "Der Deutsche Werkbund in Paris," p. 431.

42   Dr. U. Gutmann, "Zum Raum wird hier die Zeit (Zur Eröffnung der Ausstellung des Deuschen Werkbundes in Paris)," *Königsberger Hartungsche Zeitung* (May 20, 1930).

43   Herbert Bayer, "Aspects of Design of Exhibitions and Museums," *Curator: The Museum Journal* 4/3 (July 1961): p. 268.

44   Bayer, "Aspects of Design of Exhibitions and Museums," p. 268.

45   Ibid., p. 258.

46   Bayer, "Fundamentals of Exhibition Design," p. 17.

47   Mary Anne Staniszewski suggested that Bayer's chair display inspired Alfred H. Barr's hanging of chairs in the *Cubism and Abstract Art* exhibition at the MoMA. See Mary Anne Staniszewski, *The Power of Display: A History of Exhibition Installations at the Museum of Modern Art* (Cambridge, MA, 2001), pp. 75–8.

48   Bayer, "Fundamentals of Exhibition Design," pp. 22–4.

49   Maurice Merleau-Ponty highlighted that our gaze favors predetermined perceptual routes when it moves over familiar objects and does not recognize an object unless its details are perceived in a specific order. To re-establish the correct orientation between an object and ourselves, we sometimes even do so physically by tilting our head—an insight frequently deployed in avant-garde photography. Maurice Merleau-Ponty, *Phenomenology of Perception*, trans. Colin Smith (London, 1962), pp. 252–3.

50   Bayer, "Aspects of Design of Exhibitions and Museums," p. 240.

51   Giedion, "Der Deutsche Werkbund in Paris," p. 431.

52   Ibid., p. 432.

53   The possibility of the requisite having been concealed is mentioned in Jakob Gebert and Kai-Uwe Hemken, "Der Raum der Gegenwart: Die Ordnung von Apparaten und Exponaten," in *Kunst Licht Spiele: Lichtästhetik der klassischen Avantgarde*, eds. Ulrike Gartner, Kai-Uwe Hemken, and Kai Uwe Schierz (Bielefeld, 2009), p. 150.

54   Gutmann, "Zum Raum wird hier die Zeit."

55   Osborn, "Gropius und die Seinen."

56   László Moholy-Nagy, "Lichtrequisit einer elektrischen Bühne," *Die Form: Zeitschrift für gestaltende Arbeit* 5/11–12 (1930): pp. 297–8. The intended nature of the light–space modulator as a "requisite" was frequently misunderstood as a "mobile sculpture," "three-dimensional object," or "half sculpture and half machine," as Moholy-Nagy's first wife Lucia Moholy pointed out. See Lucia Moholy, *Marginalien zu Moholy-Nagy* (Krefeld, 1972), p. 41.

57   Moholy-Nagy, "Lichtrequisit einer elektrischen Bühne," p. 297.

58   Ibid., p. 299.

59   Moholy-Nagy critically examined urban life during the economic crisis in a short film made in Berlin, "Berliner Stilleben" (Berlin Still Life), 1932. The dating follows Jeanpaul Goergen's chronological listing of László Moholy-Nagy's film projects: "Filme, Projekte Vorschläge: Annotierte Filmografie 1921–1934," in *Kunst Des Lichts: László Moholy-Nagy*, eds. La Fabrica (Berlin, 2010), p. 246.

60   Oliver A.I. Botar, "Gesamtkunstwerk ohne Kunst," in *Kunst Des Lichts: László Moholy-Nagy*, eds. La Fabrica (Berlin, 2010), p. 165.

61   Lucia Moholy underlined the continuity of ideas that led to the ongoing development of the light–space modulator. The basic idea for the apparatus had already been formulated in Moholy-Nagy and Alfréd Kemény's 1922 manifesto "Dynamisch-Konstruktives Kraftsysstem." See Moholy, *Marginalien zu Moholy-Nagy*, p. 38.

62   In 1930, he produced a film capturing the light and color effects of the light–space modulator, "Lichtspiel Schwarz-Weiss-Grau" (light play black–white–grey), which was shown for the first time in Berlin, March 4, 1932. The dating follows Jeanpaul Goergen's chronological listing of László Moholy-Nagy's film projects: "Filme, Projekte, Vorschläge: Annotierte Filmografie 1921–1934," pp. 245–6.

63   Moholy-Nagy and Kemény, "Dynamisch-Konstruktives Kraftsysstem."

64   See Sigfried Giedion, José Luis Sert, and Fernand Léger, "Nine Points on Monumentality," written in 1943. Published in *Architecture, You, and Me: The Diary of a Development*, ed. Sigfried Giedion (Cambridge, MA, 1958), pp. 48–51.

65   J. Grünberg, "Die Gropius Ausstellung in Paris," *Berliner Tageblatt* (May 31, 1930).

66   Giedion, "Der Deutsche Werkbund in Paris," p. 430.

67   Gernot Böhme suggests that the progressive aestheticization of our reality is primarily a matter of material aesthetics—that is, the extensive presentation of materiality for the staging of our everyday lives. He proposes that every product has a practical value and an aesthetic or "stage" value (Inszenierungswert). The latter has become an independent value system in late capitalism that he thematizes under the heading of "aesthetic economy." See Gernot Böhme's chapter "Der Glanz des Materials: Zur Kritik der Aesthetischen Ökonomie" (The Shine of Material: On the Critique of the Aesthetic Economy), in *Atmosphäre. Essays zur neuen* Ästhetik (Frankfurt am Main, 2013), pp. 49–65.

68   Siegfried Giedion, *Space, Time and Architecture: The Growth of a New Tradition* (Cambridge, MA, 1967), p. 481.

15

Unfinished Architecture: Urban Continuity in the Age of the Complete

Nicholas Temple

INTRODUCTION

> *How can anything ever present itself truly to us since its synthesis is never*
> *completed? How could I gain the experience of the world, as I would of an*
> *individual actuating his own existence, since none of the views or perceptions I*
> *have of it can exhaust it and the horizons remain forever open?*[1]

Maurice Merleau-Ponty's meditation on incompleteness serves as an appropriate starting point in this investigation of the unfinished in building. His argument that it is impossible to gain a complete "picture" of the world, on account of the inexhaustibility of our perceptions and experiences, prompts us to question the assumption of architecture's "closure," with respect to its creative process and its experiential presence. In asserting that our horizons of experience remain forever open, Merleau-Ponty's phenomenology confronts head on the deeply embedded instrumental precepts of contemporary culture. Among the many areas where these lay authoritative claim, the transformations of the city are perhaps the most acute and visible, with their multiple systems of management and control. These modes of urban transformation, and the broader historical background of "city marking," serve as the background to this study of the life of buildings and its influence on the creative imagination of architects.

I begin, however, with a personal note: in my first job as a young academic, at Leeds Metropolitan University, I worked in a building that was left unfinished, a fact was not particularly remarkable except that it gave rise to a rich and fertile dialogue within the academic community about the role of architecture in "city making" (*cosmopoiesis*). Built in the early 1970s, the original scheme for the Brunswick Building, as it was called, centred on a semi-enclosed upper courtyard with elevated external walkways connecting different parts of the building complex. From this platform extended three linked wings, each cranked to form an incomplete polygonal-shaped piazza at lower level, with connecting ramps and steps. The resulting external space, and its surrounding ensemble of buildings,

15.1 Drawn aerial view of the Brunswick Building, Leeds, indicating the three linked wings (un-built wing shown as no. 3), and location of the terminal wall to the second wing (x).

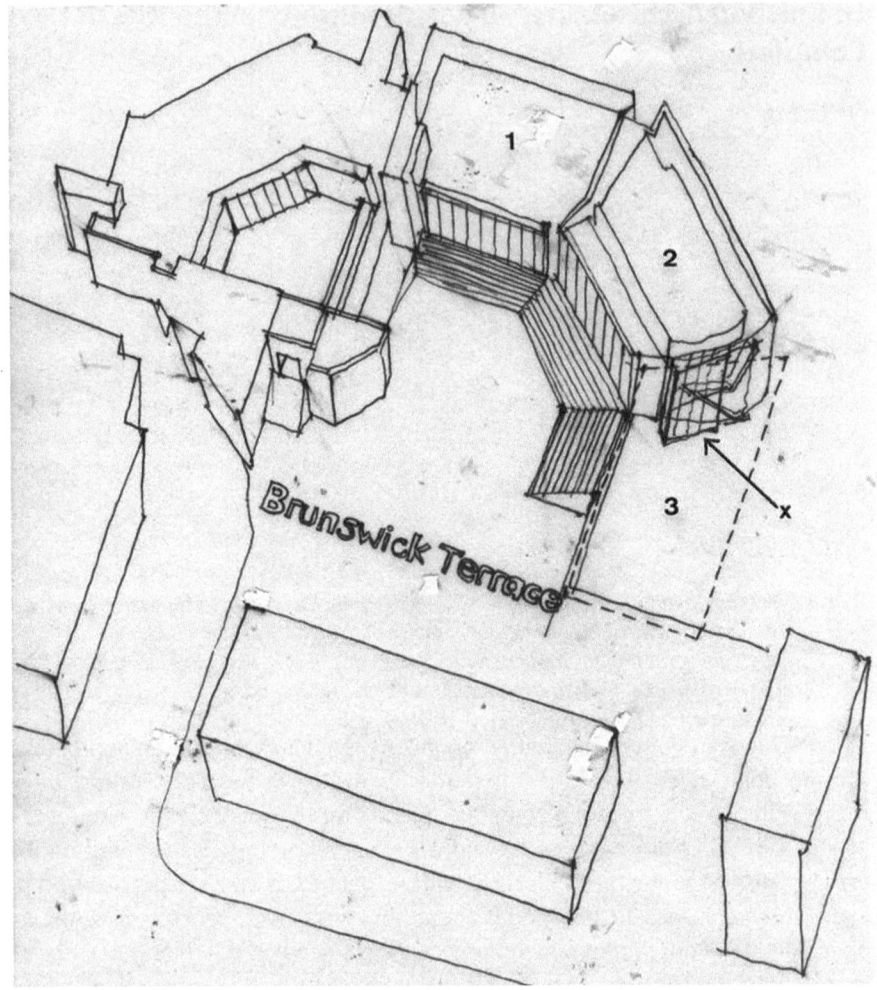

gave the Brunswick Building a distinctive civic presence, absent elsewhere in the public spaces of central Leeds.

Among the numerous omissions/revisions in the final building, the most conspicuous was the decision to omit the final wing on the east side due to budget cuts. As a consequence of this missing element, the end wall of the second wing was abruptly terminated by a makeshift fire-escape stair, with exposed reinforcing rods projecting from its in-situ concrete surface. Over time, this unfinished (and hastily assembled) end wall became an icon of the Leeds School of Architecture, resolutely standing out as a modern ruin from the banal corporate architecture that surrounded it.[2]

The studio where I taught design was located on the top floor of the second wing, which gave access to the makeshift fire-escape stair of the terminal wall. The elevated platforms of the stair served as a vantage point from where students could survey the panorama of the city. It was in this studio that I conceived and coordinated a series of urban/architectural projects using vacant urban sites

15.2   View of the "Terminal Wall" of the Brunswick Building, Leeds.

and abandoned/incomplete buildings in Leeds, many of which were visible from the vantage point of the terminal wall. These projects variously tackled the idea of architecture as an "on-going" project, drawing upon the classical precept of *renovatio urbis*, a term Peter Carl argues constitutes an essential, albeit declining, aspect of city-making:

> *The grand theme of renovatio urbis depends for its meaning upon a conception, experience, or culture of urbis that is susceptible to renewal … . [A]lthough renovatio manifests itself in the construction or restoration of buildings, streets, canals, and defense-works, and in the writing of poems or the creation or modification of laws and ceremonies …, all this activity is only a means to a more profound end. What is restored is the reciprocity between the given, historical conditions of urbis and their potential to be understood in the highest terms of goodness or beauty or lawfulness, etc.*[3]

In this investigation, I shall explore the role that unfinished architecture plays in this predisposition of cities to renewal (*renovatio urbis*), prompting us to re-evaluate how the physical performance of materials in building construction can furnish the architectural imagination through the ambiguous (open-ended) nature of suspended work.

"OPEN WORK"

The curious history of the Brunswick Building—its short life and its irresolvable state of incompletion—raises some intriguing questions about how buildings can, over time, respond to change and serve as temporal registers of both real and imagined settings. Both attributes of architecture, however, are not necessarily coincident, since adaptability in building is typically treated as exclusively a domain of spatial planning, rather than as an expression of the poetics of urban continuity through incremental alteration or adaptation. Lars Lerup ponders this relationship when he states:

> Change is normally thought of as a process that inches its way bit by bit towards the future. But if there is a real concern for the present, the future—the Arcadia—loses its relevance … .To expect fixity in the environment appears absurd against the facts of steady and personal changes among dwellers … I have previously gathered this focus and concern under the slogan: building the unfinished. Many ideas rally under this. The need for a many-sided view is one; another is the open-endedness of the environment despite our view of it as finished. All these ideas and observations suggest that we should think of the socio-material world of the dweller as largely unfinished.[4]

The slogan, "building the unfinished," seems at first a contradiction in terms and conveys a certain controversy about the status of creative work—its directives and ultimate purpose—in the contemporary world. It conflicts with the modern "teleological view of a building's progressive formal development from an anointed origin toward a final goal," and the manner in which this teleological outlook curtails the creative imagination.[5]

To probe the deeper meanings of this modern teleology I shall refer to Umberto Eco's claims of "open work" in the age of modernity.[6] Eco argues that there exists a difference between traditional and modern art forms that relate to the question of the "degree of openness"; the capacity of artworks to adapt/respond to new circumstances.[7] The implication of a "sliding scale" of openness that Eco implies is measured on the basis of how art as an "epistemological metaphor" finds expression through historical change; from the deployment of "the canon of authorised responses," which characterized the medieval world-view, with its hierarchy of fixed, preordained orders,"[8] to the "fluid state"[9] of the modern world that requires a very different creative response. As a "leitmotiv" of modernity, open work feeds our insatiable quest for individual freedom of expression. In this peculiarly modern purview, greater emphasis is given to the viewer/witness to participate in the work and ultimately in its finished state: "the author offers the interpreter, the performer, the addressee a work *to be completed*."[10] Alongside this outlook is Eco's assertion of a "flight away from the old, solid concept of necessity [characteristic of the medieval world view] and the tendency toward the ambiguous and the indeterminate … ."[11]

It is the view of this author that Eco's promulgation of "open work" puts into parenthesis the commonly held teleological perspective of modernity, with its predisposition towards "closure" rather than to openness that Eco claims. At the same time, Eco's assertion that the Baroque period reveals the first clear signs of the "indeterminacy of effect," characteristic of open work, overlooks key aspects of earlier Renaissance artistic and intellectual accomplishments. These, as I shall argue, serve as a more compelling point of reference when re-evaluating the meaning of the unfinished in modernity, particularly in regard to the concept of *renovatio urbis*.

We are reminded in this duality of "open work" of the grey stone metropolis of Fedora in Italo Calvino's *Invisible Cities*, whose centre is occupied by a metal building containing crystal globes in each room:

> *Looking into each globe, you see a blue city, the model of a different Fedora. These are the forms the city could have taken if, for one reason or another, it had not become what we see today. In every age someone, looking at Fedora as it was, imagined a way of making it the ideal city, but while he was constructing his miniature model, Fedora was already no longer the same as before, and what had been until yesterday a possible future became only a toy in a glass globe.*[12]

The perpetual state of disjunction, elucidated in Fedora, between on the one hand the reality of urbanity's relentless change (that resists momentary suspension), and on the other, speculations of future possibilities to redeem the past (the basis of ideal models), reveals how "open work" both nourishes and obstructs our imagined vision of the city. But what bearing does this dichotomy have on the role of unfinished work in perpetuating urban (cultural) renewal?

RENOVATIO URBIS AS "WORK-IN-PROGRESS"

These two modes of thinking about the city emerged as dialectically related forms of creative thought during the Renaissance, at once drawing upon an older tradition and consciously departing from it. Michel Jeanneret describes this in the following terms:

> *The humanists … sought to distinguish themselves from their predecessors, and consolidate the historical rupture that would guarantee their modernity. To this end they constructed an image … of medieval thought enslaved to rigid dogmas and immutable essences in a rigidified culture that conceived the universe as an invariable, rational, closed system … it served as a foil that allowed the sixteenth century to reject a reputedly static world vision and emancipate the mind from an order deemed reductive and inhibiting. Renaissance thinkers not only rejected this world view, they gave a positive value to change and celebrated the alteration of things and the flux of contingencies as a promise of renewal …*[13]

By standing in opposition to the medieval scholastic world-view, Renaissance humanists consciously cultivated an outlook that both drew upon the legacy of classical antiquity and sought to establish a new paradigm of knowledge based on a theoretical standpoint. Jeanneret's summary of this humanist initiative highlights a crucial feature of Renaissance culture that has a bearing on Eco's concept of open work modern and by implication on modern urban transformations, namely the new emphasis on "the flux of contingencies as a promise of renewal."

Such a mode of creativity gave impetus to the imaginative possibilities generated by *renovatio urbis*, in which the city becomes a setting where the promise of renewal is made tangible through the appropriation of existing building fabric and in the construction of new buildings and monuments. A conspicuous feature of this enterprise, as we see for example in the transformations of Rome in the early sixteenth century, was the prevalence of unfinished projects in which renewal is manifested in urban fabric as an on-going project.

This state of the unfinished, moreover, was motivated by a "determination to perpetuate the dynamics of the miraculous creative genius," thereby making creation itself "indissociable from the creating subject."[14] Such a mark of authorship in the creative process served as a metaphor for "God's cosmic creation," providing the framework for communicating the humanist project of *renovatio urbis* in architecture and the visual arts. Not surprisingly, the most fertile philosophical influence on this enterprise can be found in Plato's *Timaeus*, which saw a resurgence of interest in early sixteenth-century humanism.[15] Plato's cosmology distinguishes between "two orders of existence, the intelligible and unchanging model and the changing and visible copy … ."[16] We can see how philosophical notions of flux and alteration, that underpinned Plato's cosmology, would have informed the broader cultural milieu of change and its urban/architectural and artistic manifestations:

15.3   Detail from the fresco, the *School of Athens* (c.1509) by Raphael, showing the figure of Michelangelo "standing in" for the philosopher Heraclitus. Vatican, Stanza della Segnatura.

> *Plato's position on creation is nearer to that of Heraclitus, who alone had rejected the notion of substance underlying change and had taught the complete transformation of every form of body into every other. We are now to think of qualities which are not also "things" or substances, but transient appearances in the Receptacle.*[17]

The impact of Heraclitus' philosophy of *panta rhei* (via Plato's Receptacle) on humanist thought extended, it seems, to artistic practices and temperaments.[18] The choice of

Michelangelo to "stand in," so to speak, for Heraclitus in Raphael's *School of Athens* (c.1509) has prompted numerous speculations.[19] We can see why Raphael may have arrived at this match between the ancient philosopher, who is said to have wept incessantly (perhaps on account of his belief in the transience and impermanence of things), and the Renaissance artist noted for his restlessness and pessimism.[20] Alongside Leonardo da Vinci, Michelangelo is the most-noted Renaissance artist for leaving many of his works unfinished. One explanation, largely perpetuated by Vasari, is that his "*non finito* reflects the sublimity of his ideas, which again and again lay beyond the reach of his hand."[21] Vasari's argument, which perpetuates the idea of the unfinished as a sign of the fertile creative imagination, is based on the Neo-platonic notion that the artist's "idea was always more important than its realisation."[22] This resulted in Michelangelo's willingness to "change his design, even when it was in the course of execution."[23] We can see this, for example, in his design of Julius

II's Tomb, in which Michelangelo abandoned the original architectural framework and statues of slaves in the course of their realization.[24]

It seems that Michelangelo's resistance to following an idea to its synthesis actually underpinned his creative process, and in so doing could be construed in vaguely "Heraclitan" terms as an emulation of the divine forces of nature's perennial flux. Given these characteristics of Michelangelo's work, his appearance in the *School of Athens* as Heraclitus may not just be a conceit on the part of Raphael; the pessimistic streak in Michelangelo's character may have been interpreted by Raphael as signalling a state of mind convergent with the moral uncertainties that the early sixteenth-century Roman Catholic Church would have construed from Heraclitus' philosophy of flux; that perpetual change carries with it a burden of pessimism and doubt about what will come. Indeed, pessimism served as the flip-side to Renaissance optimism for an imminent Golden Age; that the promise of Rome as *altera Jerusalem* was always counterbalanced by its portrayal as Babylon reborn.[25]

It seems to me that this Renaissance sensibility shares some of the traits of Eco's open work, with its anticipation of modernity's multiplicity and plurality; only in the case of the art and architecture of humanism, receptiveness to change operated at the level of a dialectical relationship. This oscillated between a deeply embedded classical tradition and a newly discovered reverie towards the dynamics ("inner workings") of nature, or *natura naturans*, revealed through the combined effects of direct observation and the human creative process.[26]

In many respects unfinished, or altered, buildings in sixteenth-century Rome provided the most fertile expression of this relationship; on the one hand architecture at this time adhered to the "timeless" Platonic–Pythagorean order of harmonic proportions (exemplified in Bramante's famous parchment plan for the new St Peter's basilica). On the other hand, we see tangible evidence of building projects being subjected to (traceable) temporal change, through their abruptly suspended or appropriated states (observe, for example, the rusticated base of Bramante's abandoned Palazzo dei Tribunali along via Giulia, that was later incorporated into other buildings). In such examples the burden of material delay and decay is made clear, anticipating in the process the full consequences of Mannerist fragmentation.[27]

Jeanneret even implies that the sixteenth century was in many ways the century of the incomplete, nurtured by a "transformist sensibility."[28] We can see how this sensibility may have been influenced in part by a new sense of urgency at the beginning of the sixteenth century, a time that many believed to be auspicious. Claims, for example, of Julius II's chief spokesman, the Augustinian friar Giles of Viterbo, that this period signalled the coming of a "golden age," that could rival past golden ages in biblical history, gave impetus to the ambitious projects being undertaken by Julius II.[29] The saga of the construction of the new St Peter's Basilica, the largest and arguably most important building project in sixteenth-century Europe, serves as a powerful expression of this self-conscious age of renewal. In this project, the old and new fabric co-existed in various stages for almost a century; as the old basilica was being gradually demolished, the new was taking its place. For Federica Goffi, this co-existence created a new way of thinking about

15.4   View of the rusticated base of the uncompleted Palazzo dei Tribunali (c.1510), incorporated into later palaces along via Giulia in Rome.

the conservation of built form as a dynamic (temporal) process that enlivened invention and the imagination.[30]

It is tempting to consider the resulting mismatch between high ambitions, and the reality of unresolved and partially completed projects, as simply indicative of a creative impulse over-extending itself, in a way similar to Vasari's account of Michelangelo's "*non finito*" referred to earlier. But this implication of failure, on the part of both artist and patron, only overlooks the broader cultural context of these extraordinary initiatives; in essence the unfinished was somehow "built into" the cultural fabric of the society, as a necessary condition of human endeavour and imagination, and ultimately of collective human salvation.

Seen from the broader perspective of Rome, it is apparent that Renaissance initiatives for self-renewal, through the act of building, embraced the larger topography of the city, in which ancient ruins were being quarried for building materials, and *spolia* recycled as architectural components. It is as if Rome, with its combination of ancient ruins, building sites and large areas of semi-demolished buildings, was undergoing a continual process of transformation, in which old and new were effectively conflated. From this permanent state of transience emerged multiple "versions" of Rome in the creative imagination of architects and

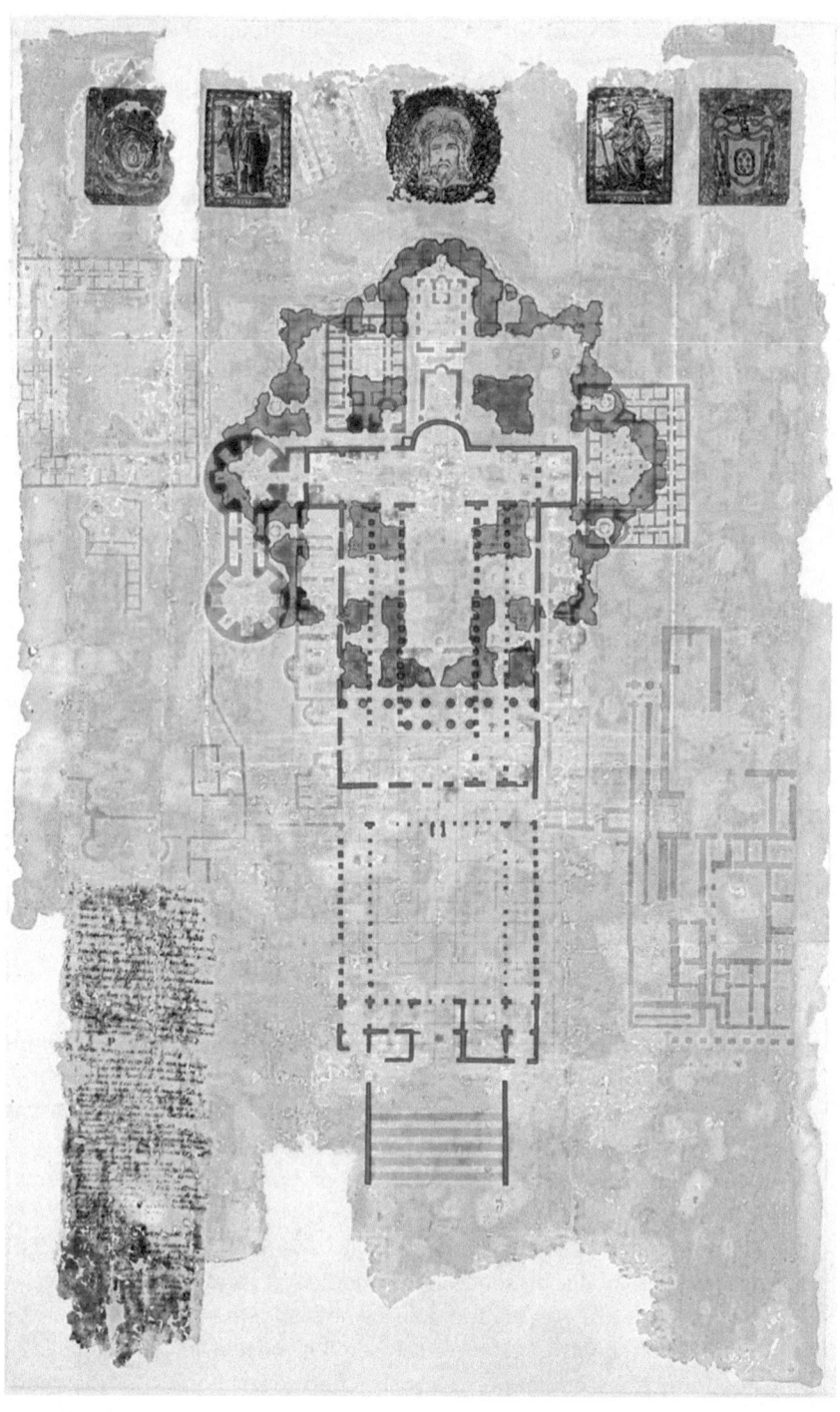

15.5 Digital restoration of Tiberio Alfarano's 1571 hybrid drawing of old and new St. Peter's Basilica.

antiquarians, each susceptible to re-invention through the fertile narratives of the city's mytho-historic past.[31]

Visible evidence of distinct stages of development in this process of transformation, as one would expect in the urban/architectural accomplishments of individual pontiffs, were rarely commemorated by the completion of buildings. Only inscriptions, dedicated to the residing or recently deceased popes, could serve as visual and epigraphic registers of these initiatives at any one period.[32]

I would argue that this historical model of architectural and artistic endeavour, whilst exceptional during the Renaissance (in terms of its scope and collective ambition), reflects deeper meanings of *renovatio urbis* that still have a bearing on the contemporary city. As Carl states, "*renovatio* appears to confront so much of the modernist project because it speaks not only to temporal renewal, but even more to the content of the temporality that is renewed."[33] It provides a "barometer" of how renewal is understood not merely as the basis of cyclical change, facilitated through commercial/corporate activity, but more fundamentally as part of the historical and cultural continuity of the city.

THE TEMPORALITY OF ARCHITECTURE

To establish a clearer understanding of the status and meaning of *renovatio urbis* in architecture today we have to go to an unlikely source—the building site. It is a significant, but largely overlooked, fact that the building site is the last remaining vestige of the processes of making and fabricating in the public realm. In place of the visible signs of crafting and producing goods and artefacts that once dominated urban life before the modern age, cities today have become little more than locations for the display and promotion of prefabricated and virtual products in shops, offices and public spaces.

With the compressed timescale of modern building construction, on account of the increasing costs of labour and materials, and the demands for more effective health and safety regulations in the building industry, these sites of transformation in the city have become effectively closed off from public gaze. Against the overarching impression of homogeneity and completeness, which the contemporary corporate city presents to the general public, the "messy" processes of construction (through the exposure of building carcasses) seem strangely incongruous.

The problematic status of the building site today contrasts with its enduring and legitimate public presence in the Middle Ages and the Renaissance, where the act of building was in every sense an acknowledgement of architecture's perennial unfinished or adaptive state. We can see this most vividly in the building sites of cathedrals, whose duration of construction (over many centuries and through successive episcopates), was characterized by expanses of semi-permanent scaffolding and assembled piles of cut stone and formwork. These sites were almost certainly a more familiar feature of medieval urban life than finished places of worship, an issue that is often ignored by architectural historians. An indication of the consequences of this accumulated history of building sites is highlighted in the following statement:

> In modernity building sites do not last for long, but in the pre-modern period
> the typical monumental "building" was in reality a building site for so long that
> sometimes this condition remained as a palimpsest—for example, in naming, as
> in the streets around the north, east and south sides of Florence Cathedral, whose
> substructures were so long in the making that the area came to be called the Via
> dei Fondamenti.[34]

A key factor in these hugely ambitious projects was the capacity of building work to be conceived and represented through different modalities of time. Marvin Trachtenberg examines this aspect of building:

> The opposition of architecture and time is generally so strong in modernity
> … that it tends to be difficult to see concretely beyond it to a world of non-
> chronicidal architectural temporality. Yet my analysis raises the possibility that
> the architecture–time relationship might be alternatively construed, not as
> merely "neutral" but in altogether positive terms. This would require that time be
> seen other than as a malignant force practically or intellectually.[35]

Trachtenberg's argument of a "chronicidal" form of temporality in modernity derives from his argument of a historical shift from "building-in-time" to building-outside-time. This transformation took place as a result of changes in the relationship between two modes of temporality within architecture itself: the duration of a building's construction and its use/inhabitation (or "lifeworld"). Trachtenberg highlights the complex overlaps between both, in which the slowness of construction in pre-modern times meant that the lifeworld was much more intrusive in the building process. This is demonstrated, for example, in the on-going alterations to the new St Peter's Basilica during its construction. In the case of modern architecture, however, "the velocities of both architectural making and of the lifeworld not only are in relative conformity but also the speed of construction is also so great that lifeworld conditions usually do not have time to change enough to affect the project much during the execution of the final design."[36] The absence of visible evidence of changes in a building's design, in the process of its construction, is what Trachtenberg describes as "building-outside-time, a closed teleological approach that has its origins in Albertian theory of architecture.[37] Stephen Parcell provides a useful spatial (Janus-faced) model when considering the question of the role of architecture in acknowledging a past and anticipating a future:

> We normally presume that a work emerges from a world that has preceded it.
> Conceived as the off-spring of a specific author in a specific time and place, it
> would seem to be the end product of intentions and historical forces. … However,
> even after an intricate historical background has been established, the work itself
> retains a degree of independence.[38]

Parcell argues that the resulting "gap between things and ideas enables a work to be disengaged from the world behind (its history). Protected by a conceptual moat, it is free to engage in other discourses opening up in front (its fictions), introduced and witnessed by the architectural performer."[39] In other words, the work is situated at an interface between, on the one hand, a pre-existing context and, on the other, future possibilities. Parcell's thesis may well have drawn a literary analogy from Paul Ricoeur's concept of "Distanciation," as it pertains to the tension between the "condition" of understanding a written text (by virtue of its autonomy and remoteness) and the need to "conquer" it by hermeneutical means.[40]

In this "bifrons" model of the temporality of architecture the face looking backwards is masked, by virtue of the "conceptual moat" that Parcell describes. This situation is perpetuated by the standpoint that "contemporary historiography [and we could say of contemporary culture in general] gives precedence to one of the illusions of consciousness, that the perspective of our own historical moment must be autonomous."[41] The resulting asymmetry in the temporal understanding of architecture is, I believe, one of the most critical challenges facing architects today, since it ignores the fundamental role of memory in the creative imagination.[42] Parcell warns of the potential dangers of such disengagement of the work from its historical background, in the way it can lead to "a sterile object floating in a universal kit of parts."[43]

It seems, however, incontestable that much architecture today "revels" in this autonomy and detachment, by asserting its capacity to operate freely as a monologue with itself, rather than acting in dialogue with the past through acknowledgement (and appropriation) of a prevailing set of cultural practices (what we broadly call tradition).

It is in the context of Parcell's particular interpretation of the temporal dimensions of architecture, that we can begin to understand more clearly the status of Eco's model of "open work" in the contemporary city. In the continuing and relentless search for new innovations, which characterizes our technologically driven society, open work has become a largely opaque process concealed behind the visual and bureaucratic layers of systems and organizations. The role of architecture, however, in this strategy of concealment remains problematic, as we have seen in the context of the archaic presence of the building site. In spite of Trachtenberg's assertion of the impact of the speed of construction today on the disappearance of visible signs of "lifeworld" changes/adaptions in building (when compared to pre-modern times), the place-specific nature of building construction, and its enduring spatial and temporal presence in the city, serve as persistent reminders of architecture's role as an embodiment of urban (and cultural) continuity.

In this investigation I have argued the importance of unfinished work in the collective memory of the city, and how this memory of previous or unresolved undertakings provides an essential ground for projecting future possibilities of architecture, through the material imagination. The example of the Brunswick Building, examined at the beginning of this chapter, demonstrates how unintended suspension of building work, and its architectural consequences, can act as a catalyst for creative reflection. Adapting a literary reference taken from Ricoeur, "everyday reality is metamorphized by means of what we could call the imaginative variations that [unfinished building] works on the real."[44]

BIBLIOGRAPHY

Bialostocki, J., "The Renaissance Concept of Nature and Antiquity," in *The Renaissance and Mannerism: Studies in Western Art. Acts of the 20th International Congress of the History of Art* (1963): pp. 19–30.

Burroughs, Charles, *From Signs to Design: Environmental Process and Reform in Early Renaissance Rome* (Cambridge, MA: MIT Press, 1990).

Calvino, Italo, *Invisible Cities* (New York: Harcourt Brace Jovanovich Inc., 1974).

Carl, Peter, "Renovatio and the Howling Void," in Mohsen Mostafavi and Homa Fardjadi, *Delayed Space* (New York: Princeton Architectural Press, 2001), 18–37.

Cornford, Francis, M., *Plato's Cosmology: The Timaeus of Plato* (Indianapolis, IN: Hackett Publishing Company, 1997).

Eco, Umberto, *The Open Work* (Cambridge, MA: Harvard University Press, 1989).

Goffi, Federica, *Time Matter(s): Invention and Re-Imagination in Built Conservation: The Unfinished Drawing and Building of St. Peter's, The Vatican* (Farnham, Surrey: Ashgate, 2013).

Hankins, James, *Plato in the Italian Renaissance* (Leiden: E.J. Brill, 1991).

Jeanneret, Michel, *Perpetual Motion: Transforming Shapes in the Renaissance from da Vinci to Montaigne* (Baltimore, MD: Johns Hopkins University Press, 2001).

Kahn, Charles H., *The Art and Thought of Heraclitus: An Edition of the Fragments with Translation and Commentary* (Cambridge: Cambridge University Press, 1979).

Kiang, Dawson, "Bramante's 'Heraclitus and Democritus': The Frieze," *Zeitschrift Kunstgeschichte*, 51/2 (1988): pp. 262–8.

Lerup, Lars, *Building the Unfinished: Architecture and Human Action* (Beverly Hills, CA: Sage Publications, 1977).

Lutz, Cora E., "Democritus and Heraclitus," *The Classical Journal* 49/7 (April, 1954): pp. 309–14.

Hall, Marcia (ed.), *Raphael's School of Athens* (Cambridge: Cambridge University Press, 1997).

Merleau-Ponty, M., *Phénoménologie de la Perception* (Paris: Gallimard, 1945).

Pallasmaa, Juhani, *The Eyes of the Skin: Architecture and the Senses* (West Sussex: John Wiley & Sons Ltd., 2005).

Parcell, Stephen, "The World in Front of the Work," *Journal of Architectural Education* 46/4 (May, 1993): pp. 249–59.

Ricoeur, Paul, *Essays on Biblical Interpretation* (London: SPCK, 1981).

Ricoeur, Paul, "The Hermeneutical Function of Distanciation," *Philosophy Today* 17.2 (1973): pp. 129–41.

Schulz, Juergen, "Michelangelo's Unfinished Works," *The Art Bulletin* 57/3 (September 1975): pp. 366–73.

Temple, Nicholas, "Mannerism," in *The Oxford Dictionary of Christian Art & Architecture*, edited by Peter and Linda Murray and Tom Devonshire Jones (Oxford: Oxford University Press, 2013), pp. 344–7.

Temple, Nicholas, renovatio urbis: *Architecture, Urbanism and Ceremony in the Rome of Julius II* (London: Routledge, 2011).

250   THE MATERIAL IMAGINATION

Trachtenberg, Marvin, *Building-in-Time: From Giotto to Alberti and Modern Oblivion* (New Haven, CT: Yale University Press, 2010).

Viterbo, Egidio da, "Fulfillment of the Christian Golden Age under Pope Julius II: A Text of a Discourse of Giles of Viterbo 1507," edited and commentary by John W. O'Malley, *Traditio: Studies in Ancient and Medieval History, Thought and Religion* XXV (1969): pp. 265–338.

Zorach, Rebecca, *The Virtual Tourist in Renaissance Rome: Printing and Collecting the Speculum Romanae Magnificentiae* (Chicago, IL: University of Chicago Library, 2008)

NOTES

1   M. Merleau-Ponty, *Phénoménologie de la Perception* (Paris, 1945), p. 381. Quoted in translation in Umberto Eco, *The Open Work* (Cambridge, MA, 1989), p. 17.

2   In 2009 the site of the Brunswick Building was sold by the university to a developer, and the building—by then abandoned and the school of architecture relocated to new premises—was demolished.

3   Peter Carl, "Renovatio and the Howling Void," in Mohsen Mostafavi and Homa Fardjadi, *Delayed Space* (New York, 2001), pp. 18–37, 19.

4   Lars Lerup, *Building the Unfinished: Architecture and Human Action* (Beverly Hills, CA, 1977), pp. 142–3.

5   Marvin Trachtenberg, *Building-in-Time: From Giotto to Alberti and Modern Oblivion* (New Haven, CT, 2010), p. XXIII.

6   See note 1.

7   Eco, *The Open Work*, p. xii.

8   Ibid., p. 13.

9   Ibid., p. 7.

10   Ibid., p. 19.

11   Ibid., p. 17.

12   Italo Calvino, *Invisible Cities* (New York, 1974), p. 32.

13   Michel Jeanneret, *Perpetual Motion: Transforming Shapes in the Renaissance from da Vinci to Montaigne* (Baltimore, MD, 2001), p. 3.

14   Ibid., pp. 2–3.

15   James Hankins, *Plato in the Italian Renaissance* (Leiden, 1991), pp. 161–263.

16   Francis M. Cornford, *Plato's Cosmology: The Timaeus of Plato* (Indianapolis, IN, 1997), p. 177.

17   Ibid., p. 178.

18   Jeanneret, *Perpetual Motion*, p. 29. For an examination of Heraclitus' philosophy of flux see Charles H. Kahn, *The Art and Thought of Heraclitus: An Edition of the Fragments with Translation and Commentary* (Cambridge, 1979), pp. 147–53. As is the case elsewhere, the Renaissance was especially partial to demonstrating dialectical relationships between ideas—through allegory or symbolic programmes—as we see for example in Bramante's relief of Heraclitus and Democritus; Dawson Kiang, "Bramante's 'Heraclitus and Democritus': The Frieze," *Zeitschrift Kunstgeschichte*, 51/2 (1988): pp. 262–8.

19   Ingrid Rowland, "The Intellectual Background of the *School of Athens*," in *Raphael's School of Athens*, ed. M. Hall (Cambridge, 1997), pp. 131–70, pp. 57–8.

20   On Heraclitus as the weeping philosopher see Cora E. Lutz, "Democritus and Heraclitus," *The Classical Journal*, 49/7 (April, 1954): pp. 309–14.

21   Quoted in Juergen Schulz, "Michelangelo's Unfinished Works," *The Art Bulletin*, 57/3 (September 1975): pp. 366–73.

22   Ibid., p. 366.

23   Ibid., p. 373.

24   Ibid., p. 368.

25   This twofold model is largely drawn from St Augustine. Nicholas Temple, renovatio urbis: *Architecture, Urbanism and Ceremony in the Rome of Julius II* (London, 2011), pp. 2, 43–9.

26   J. Bialostocki, "The Renaissance Concept of Nature and Antiquity," in *The Renaissance and Mannerism: Studies in Western Art. Acts of the 20th International Congress of the History of Art* (1963), pp. 19–30.

27   For an account of the key characteristics of Mannerism see my entry in *The Oxford Dictionary of Christian Art & Architecture*, ed. Peter and Linda Murray and Tom Devonshire Jones (Oxford, 2013), pp. 344–7.

28   Jeanneret, *Perpetual Motion*, pp. 1–7.

29   Egidio da Viterbo, "Fulfillment of the Christian Golden Age under Pope Julius II: A Text of a Discourse of Giles of Viterbo 1507," edited and commentary by John W. O'Malley, *Traditio: Studies in Ancient and Medieval History, Thought and Religion*, XXV (1969): pp. 265–338.

30   Federica Goffi, *Time Matter(s): Invention and Re-Imagination in Built Conservation: The Unfinished Drawing and Building of St. Peter's, The Vatican* (Farnham, 2013).

31   These "versions" of Rome are most palpably expressed in the proliferation of prints of the city's topography in the sixteenth century. See Rebecca Zorach, *The Virtual Tourist in Renaissance Rome: Printing and Collecting the Speculum Romanae Magnificentiae* (Chicago, IL, 2008).

32   Charles Burroughs argues that this prevalence of commemorative inscriptions in fifteenth-century Rome gave urban topography a certain "para-textual efficacy." Charles Burroughs, *From Signs to Design: Environmental Process and Reform in Early Renaissance Rome* (Cambridge, MA, 1990), p. 10.

33   Carl, "Renovatio and the Howling Void," p. 21.

34   Trachtenberg, *Building-in-Time*, p. XII.

35   Ibid., p. 14.

36   Ibid., p. XIV.

37   Ibid., pp. 70–101. Trachtenberg emphasizes the exceptional nature of Alberti's a-temporal theory of architecture in relation to the prevailing practice of architecture in fifteenth-century Italy, suggesting its prophetic nature in anticipating things to come.

38   Stephen Parcell, "The World in Front of the Work," *Journal of Architectural Education*, 46/4 (May, 1993): pp. 249–59, p. 250.

39   Ibid., p. 250.

40   Paul Ricoeur, "The Hermeneutical Function of Distanciation," *Philosophy Today*, 17.2 (1973): pp. 129–41, p. 130.

41   Paul Ricoeur, *Essays on Biblical Interpretation* (London, 1981), p. 27 (from the Introduction by Lewis S. Mudge).

42   Juhani Pallasmaa, *The Eyes of the Skin: Architecture and the Senses* (Chichester, West Sussex, 2005), pp. 67–70.

43   Parcell, "The World in Front of the Work," p. 250.

44   Ricoeur, "The Hermeneutical Function of Distanciation", p. 141.

Index

Note: illustrations are indicated by page numbers in bold.